Oxidation Ditches
In Wastewater Treatment

Oxidation Ditches
In Wastewater Treatment

By

Mikkel G. Mandt
Bruce A. Bell

ANN ARBOR SCIENCE
THE BUTTERWORTH GROUP

Library of Congress Catalog Card Number 82-70700
ISBN 0-250-40430-3

Manufactured in the United States of America
All Rights Reserved

Butterworths, Ltd., Borough Green, Sevenoaks
Kent TN15 8PH, England

PREFACE

The use of oxidation ditches for wastewater treatment is not new. In recent years, however, there has been considerable renewed interest in the use of oxidation ditches. Much of the work has been done by the various manufacturers that offer proprietary oxidation ditch systems. Partially as a result of this proprietary interest, it is difficult to find, in one place, information related to the design and operation of oxidation ditches. This book was prepared in response to the need for a compilation of such information.

In this book, we have attempted to provide the reader with a mix of theoretical and practical information related to oxidation ditches. The theoretical kinetic basis for design is combined with empirical data and equations to allow the designer to find all required information in one place. Numerous examples are included to guide the reader in the use of the information presented. Data, kinetic constants and information such as labor requirements are presented to provide a single source for such information.

Mikkel G. Mandt
Bruce A. Bell

ACKNOWLEDGMENTS

We would like to express our appreciation to those individuals and manufacturers, too numerous to name, who contributed to this effort. We also would like to thank Ms. Dew Vaughn for her patience in typing and retyping the manuscript.

Mandt   **Bell**

Mikkel G. Mandt is a graduate of Ohio State University with Bachelor's and Master's degrees in Sanitary Engineering. He is a registered Professional Engineer, and has consulted widely in North America and throughout the world. Mr. Mandt has been granted more than 40 patents worldwide in the areas of biological process, aeration, disinfection and flocculation.

At Kimberly-Clark Corporation, he conducted much of the original research and development on the ejector-powered oxidation ditch, leading to wide use of that technology for treatment of pulp and paper mill wastes. Later, he applied the technology to municipal wastewaters.

Mr. Mandt served as Director of Research and Development for the Penberthy Division of Houdaille Industries, Inc., and later as President of the Pentech Division of Houdaille Industries. During his tenure with Houdaille, he was responsible for the development of many innovative treatment techniques, including further developments and innovations in oxidation ditches.

Mr. Mandt was responsible for bringing advanced treatment techniques into Japan, Asia, Australia, Canada, Mexico, South America and Europe. He has contributed to the design of more than 100 treatment plants, either directly or indirectly.

Mr. Mandt serves on the American Society of Civil Engineers Subcommittee for Oxygen Transfer. He is a member of the Water Pollution Control Federation and the Technical Association of the Pulp and Paper Industry. He has published numerous technical articles on wastewater treatment.

Bruce A. Bell is Professor of Engineering at the George Washington University, Washington, DC, where he directs the program in Environmental Engineering. He received his Bachelor's degree in Civil Engineering, and Master's and PhD in Environmental Engineering from New York University.

Dr. Bell has both academic and professional experience. He previously taught at New York University and managed a technical marketing group for a major manufacturer of wastewater treatment equipment. Dr. Bell has also been vice-president and director of environmental engineering design for an environmental engineering firm. He has extensive experience in the design of wastewater treatment facilities. Dr. Bell is currently active as a consultant to government and industry. Author of numerous publications, Dr. Bell is a registered Professional Engineer and a Diplomate of the American Academy of Environmental Engineers.

To our wives, Jan and Catherine, with thanks and love

CONTENTS

CHAPTER 1

HISTORY, DEFINITION AND DEVELOPMENT

BACKGROUND

The continuous loop reactor, CLR (commonly called the oxidation ditch), was developed during the 1950s at the Research Institute for Public Health Engineering (TNO) in the Netherlands. The first oxidation ditch plant was placed in service in 1954 at Voorshopen, Holland. The plant was designed by Dr. A. Pasveer of TNO to service a population equivalent of 360 persons and was of the intermittent flow type, in which the ditch also serves as the final clarifier. As a result of Dr. Pasveer's effort, the process has also become known as the Pasveer Ditch [1-3].

Since the original application in Holland, the oxidation ditch has become a significant treatment technique in Europe, Australia, South Africa and North America. By 1976 there were well over 500 plants in North America alone. In recent years the rate of application of oxidation ditch plants has increased dramatically.

The oxidation ditch is a variant of the activated sludge process in which a continuous loop reactor is used to provide a bioreactor in which mixed liquor is recirculated continuously through a closed aeration channel. The oxidation ditch is generally employed in the extended aeration mode, i.e., long hydraulic and solids retention times and low organic loading rates. A directionally controlled aeration and mixing device is used to impart horizontal velocity to the reactor contents and thus circulate mixed liquor around the closed aeration channel.

The original Pasveer Ditch was developed by the TNO in order to provide reliable, inexpensive treatment systems for small communities. The original applications were nothing more than dirt tracts with sloped sides and sod siding for stabilization. The liquid operating depth was typically one meter (Figure 1.1). Oxygenation, propulsion and mixing were provided by a horizontal surface rotor (Kesner Brush). The Kesner Brush had previous application to rectangular basins, providing surface aeration in a pattern similar to spiral roll aeration. Pasveer Ditches typically operated on a fill and draw basis. Raw sewage was introduced into the ditch during daytime, aerated and agitated. At night the surface rotor was turned off, solids were allowed to settle, and clear supernatant was drawn off the surface of the oxidation ditch. The first plant constructed was a remarkable success, achieving 97% biological oxygen demand (BOD) removal. Later Dr. Pasveer experimented with ammonia and nitrate removal after observing underaerated systems.

Continuous flow Pasveer Ditches were developed in response to increased flow and organic loads. The early 1960s saw oxidation ditches spread throughout Europe. In 1964 the Canadian (Ontario) Water Resources Committee investigated oxidation ditches [4] and concluded: "It may be concluded on the basis of acquired information the oxidation ditch treatment system is rather inexpensive to construct and simple to operate, and that it produces an acceptable effluent consistently."

In 1967 LeCompt and Mandt [5,6] developed the first application of a submerged aeration and propulsion system to oxidation ditches. Banks of ejectors powered by recirculated mixed liquor and compressed air were aligned to discharge along the flow path, providing necessary oxygenation and propulsion, as shown in Figure 1.2. This process later became known as the Jet Aeration Channel (JAC).

In 1968 Dutch engineers working for Dwars, Heederik, and Verhay, Ltd. (DHV) used slow-speed surface turbines in a folded CLR. The slow-speed surface turbine was placed at the end of a center baffle, and the radial component of flow from the slow-speed surface aerator was harnessed to provide the propulsion for the oxidation ditch. This process became know as the Carrousel process (Figure 1.3).

Since the early developments of Pasveer, LeCompt, Mandt and DHV, numerous process and mechanical modifications have been applied to oxidation channels. The flexibility and range of application of the oxidation ditch from the first dirt track have expanded tremendously. The process is now firmly established and accepted as a treatment technique. In 1978 the United States Environmental Protection Agency (U.S. EPA) published "A Comparison of Oxidation Plants to Competing Processes for Secondary Advanced Treatment of Municipal Wastes" [7]. Figure 1.4, taken from that publication, shows the dramatic growth of municipal oxidation ditch plant installations in the United States. That same publication concluded: "results of this study showed that oxidation ditch plants are capable of consistently

Figure 1.1. Pasveer oxidation channel (courtesy of Lakeside Equipment Corporation).

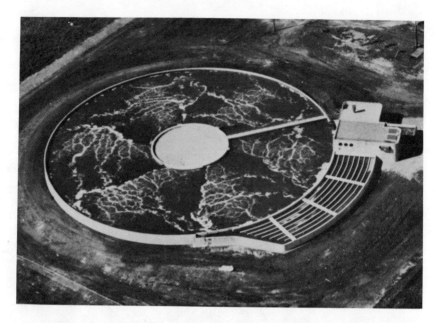

Figure 1.2. Jet aeration channel (courtesy of Fluidyne Corporation).

achieving high levels of BOD_5 and TSS removals with minimal operation. In addition, cost data . . . indicate that oxidation ditch plants are economically competitive with other processes in the flow range of 0.1 to 10-MGD." The same publication goes on to state: "All information indicates the current trend is toward increasing numbers of oxidation ditch plants, especially in the size range up to 1.5 MGD." Only a few larger plants are presently in operation. There are only a few states with no actual or planned oxidation ditch plant installations.

Increased installation of oxidation ditch plants was related to some or all of the following considerations:

1. Construction costs are equal to or less than competitive treatment processes.
2. Plants require minimum mechanical equipment.
3. Plants appear to perform reasonably well even with minimum operational attention, primarily due to conservative design.
4. Waste sludge is relatively nuisance-free and is readily disposed of at most plants.
5. Plants generally do not generate odors even under poor operating conditions.

According to W. F. Ettlich [7] : "Discussions with plant operators, public works officials, consulting engineers and others that have had direct experience of oxidation ditch plants indicate a high level of satisfaction with,

Figure 1.3. Carrousel system (courtesy of Envirotech Corporation).

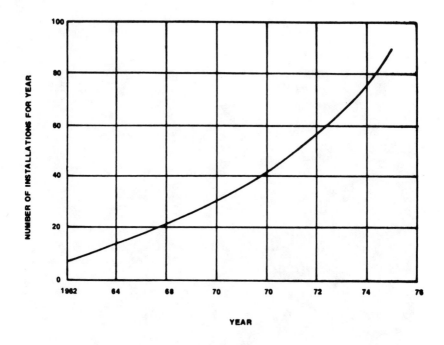

Figure 1.4. Growth of municipal oxidation ditch plants in the U.S.

and acceptance of, oxidation ditch plants. There are a few exceptions and there are a few oxidation ditch plants that have been removed from service for various operational reasons, but these are a small minority.

There are no apparent factors that would cause the rate of application of oxidation ditch plants in the United States to decrease in the near future."

Surprisingly, with the growth in oxidation ditch application there has not been a concomitant growth in public and technical information to provide a fundamental understanding of the process, nor have there been solid, rational procedures developed for design and selection of oxidation ditches. This text represents an attempt to help fill that void.

TYPICAL CHARACTERISTICS

Though oxidation ditches have a variety of modifications, most applications to municipal wastewater treatment are similar. In a typical municipal application, no primary clarification precedes the oxidation ditch. Influent screening, comminution and grit removal are recommended. The aeration basin is typically sized to operate in the extended aeration mode with hydraulic retention times (HRT) in the range of 15-30 hr and solid retention time (SRT) exceeding 15 days. HRT and SRT are defined by Equations 1.1 and 1.2:

$$HRT = \frac{\text{volume of bioreactor}}{\text{average daily flow}} \tag{1.1}$$

$$SRT = \frac{\text{solids undergoing aeration}}{\text{solids wasted per day}} \tag{1.2}$$

Directionally controlled aeration and mixing devices, such as aerating jets or mechanical rotor aerators, are employed. The rotor aerators are of the horizontal brush, cage, or disc type designed specifically for oxidation ditch plants. Secondary clarifiers are normally provided. The secondary clarifiers may be sized somewhat conservatively to provide for light, dispersed growth floc sometimes produced in oxidation channels. Table 1.1 is a summary of design parameters for typical municipal oxidation ditch plants.

The geometry of the closed loop channel varies considerably, although the oval configuration is most common. In South Africa, multiple interconnected channels are common; in Europe, folded and convoluted channels are common; in Australia and in the United States oval channels are most common, though circular channels are gaining favor. Water depths of up to 7 m have been used. Most ditches, though, are restricted to 1-2 m water depth. Many ditches use trapezoidal cross sections with 45° sloping sidewalls. Deep channels use vertical sidewalls, normally built of reinforced concrete. Gunnite, asphalt and thin membranes have been used for channel bottoms and slope sides.

Table1.1. Typical Municipal CLR Design Parameters

HRT, hr	15-40
SRT, days	>15
MLSS (mixed liquor suspended solids), mg/l	2000-6000
F/M (food/microorganism ratio)	0.05-0.02
Return Activated Sludge Rate, % of influent flow	100
BOD Removal Efficiency, %	94-98+
TSS Removal Efficiency, %	90-95+

REFERENCES

1. Pasveer, A. "A Contribution to the Development in Activated Sludge Treatment," *J. Proc. Inst. Sew. Purif.*, 4(1959).
2. Pasveer, A. "New Developments in the Application of Kessener Brushes (Aeration Rotors) in the Activated Sludge Treatment of Trade Waste Waters," Proceedings of the Second Symposium on the Treatment of Waste Waters (1960).
3. Barrs, J. K. "The Use of Oxidation Ditches for the Treatment of Sewage from Small Communities," Bulletin, World Health Organization (1962).
4. Guillaume, F. "Evaluation of the Oxidation Ditches as a Means of Wastewater Treatment in Ontario," The Ontario Water Resources Commission, Division of Research, Publication No. 6 (July, 1964).
5. LeCompte, A. R., and M. G. Mandt. "A Case Study of Effluent Treatment by Channel Aeration Propelled and Oxygenated with Ejectors," Paper Presented at the 44th Annual Conference of WPCF (Oct., 1971).
6. LeCompte, A. R. "A Pasveer Ditch American Jet Style," Proceedings of the 29th Purdue Industrial Waste Conference (1974).
7. Ettlich, W. F. "A Comparison of Oxidation Ditch Plants to Competing Processes for Secondary and Advanced Treatment of Municipal Wastes," U.S. EPA, 600/2-78-051 (March, 1978).

CHAPTER 2

PROCESS KINETICS

The primary goal for any biological treatment system, including continuous loop reactors (CLRs), is the removal of organic solids with the concomitant reduction in oxygen demand. These organic solids may be settleable, colloidal or dissolved. We rely on a physical process—sedimentation—to remove settleable solids. Colloidal and dissolved organic solids are removed through the use of biochemical processes. The secondary goal of treatment in CLRs is often the conversion of ammonia to nitrate for reduction of ultimate oxygen demand and/or removal of nitrogen for nutrient control. These topics are discussed fully in Chapter 4.

Since the environmental engineer, in contrast to the pharmaceutical industry, works with a nonsterile environment, control of biochemical processes is accomplished through control of the environment within the bioreactor to favor the growth of the desired organisms. Environmental control of bioreactors is used despite the difficulties often encountered in predicting environmental conditions and the response of the biological system to changes in environmental conditions.

PROCESS MICROBIOLOGY

In order to understand the biokinetics of continuous loop reactors, it is necessary first to look at the organisms of importance in biological systems. The microorganisms of interest to us in biological waste treatment systems belong to the kingdom of Protista. Protists are unicellular or multicellular organisms with no tissue differentiation. Protists may, for the purposes of the environmental engineer, be classified conveniently by their carbon and energy sources, or by their ability to utilize molecular oxygen.

9

Classification of Microorganisms

Organisms which utilize organic compounds for their carbon and energy sources are heterotrophic. Heterotrophic organisms are responsible for the conversion of organic matter to carbon dioxide as well as denitrification in biological treatment systems. Autotrophic organisms are those whose carbon source is inorganic, usually bicarbonate ions or carbon dioxide. Autotrophs may be further classified into photosynthetic organisms whose energy source is light, and chemosynthetic organisms whose energy is derived from the oxidation-reduction of inorganic compounds. Chemosynthetic autotrophs that use ammonia as their energy source are responsible for nitrification in continuous loop reactors [1].

Organisms may also be classified by their ability to utilize molecular oxygen. Strict (obligate) aerobes require molecular (dissolved) oxygen. Strict (obligate) anaerobes utilize chemically bound oxygen, such as may be found in sulfates. Molecular oxygen is toxic to obligate anaerobes. Facultative organisms can utilize either free or chemically bound oxygen [1].

The majority of treatment in continuous loop reactors is accomplished by bacteria. Bacteria are single-celled protists that utilize soluble food, require moisture and reproduce by binary fission. They range in size from about 0.5 to 15 μm. Bacterial cells are 80 to 90% water and the dry portion is approximately 90% organic. The organic fraction may be approximated to be $C_5 H_7 O_2 N$. The inorganic fraction consists of $P_2 O_5$ (50%), SO_3, $Na_2 O$, and CaO, plus trace nutrients such as iron and copper [2].

Fungi, which are normally defined by the environmental engineer as multicellular, heterotrophic protists, are also of importance in biological treatment. Fungi are obligate aerobes that prefer low pH (optimum of approximately 5.6) and have a low nitrogen requirement. In suspended growth reactors such as continuous loop reactors, fungi often cause problems in solids separation. They form filamentous growths which are settled poorly. Fungi are a particular problem in the treatment of low-pH or nitrogen-deficient wastes since they compete favorably with bacteria under these conditions [2].

Also of importance in biological treatment systems are unicellular protists such as protozoans, and multicellular organisms such as rotifers. These organisms perform an effluent "polishing" role by consuming small, colloidal organic particles and individual bacterial cells. Thus, these organisms serve to reduce turbidity in biological treatment system effluents [2].

METABOLISM/ENERGY

Microorganisms require a great deal of energy. It has been shown that 1 g of *Escherichia coli* requires approximately 0.6 W for cell maintenance and reproduction [3]. Thus, it can be shown that 170 lb of *E. coli* will require approximately 956,000 kcal/day, compared with a 170-lb man who requires from 2500–3000 kcal/day. The microorganisms derive their energy from oxidation-reduction reactions.

Energy Reactions

The overall reaction involved in heterotrophic metabolism may be described by:

$$\text{organics} + O_2 + \text{cells} \longrightarrow CO_2 + H_2 + \text{more cells} + \text{energy} \qquad (2.1)$$

The organic material is utilized for both energy and cell synthesis. The oxidation-reduction reactions involve the transfer of electrons from a reduced substance (the electron donor) to an oxidizing substance (the electron acceptor). We normally think of the electron donor as food. Heterotrophic metabolism utilizes organic electron donors; autotrophic metabolism uses inorganic electron donors. In aerobic systems, oxygen is the terminal electron acceptor. In anoxic systems, the terminal electron acceptor is nitrite/nitrate, while CO_2 and SO_4 serve as the terminal electron acceptor in anaerobic systems. The terminal electron acceptor determines the amount of energy which is available from the "food."

McCarty [4] has developed a stoichiometric approach to energy production and cell synthesis in biological processes. Table 2.1 presents the oxidation-reduction reactions developed by McCarty.

Example 2.1 Utilizing the equations in Table 2.1, the energy yield from the metabolism of domestic sewage may be determined for aerobic, anoxic and anaerobic conditions.

Taking the composition of domestic wastewater to be $C_{10}H_{19}O_3N$ and utilizing Equation 7 for the electron donor (food) and Equation 3 for aerobic conditions, we can write the following half reactions:

Donor:

$$(1/50)C_{10}H_{19}O_3N + (9/25)H_2O + (9/50)CO_2 + (1/50)NH_4^+ +$$

$$(1/50)HCO_3^- + H^+ + e^- \quad \Delta G^\circ = +7.6 \text{ kcal/}e^- \qquad (2.2)$$

Acceptors:

$$(1/4)O_2 + H^+ + e^- = (1/2)H_2O \qquad \Delta G^\circ = -18.675 \text{ kcal/}e^- \qquad (2.3)$$

Table 2.1. Half Reactions for Bacterial Systems [4]

Reaction Number	Half Reaction	$\Delta G°$ (W)[a] kcal per Electron Equivalent
	Reactions for bacterial cell synthesis (R_c)	
	Ammonia as nitrogen source:	
1	$1/5\ CO_2 + 1/20\ HCO_3^- + 1/20\ NH_4^+ + H^+ + e^- = 1/20\ C_5H_7O_2N + 9/20\ H_2O$	
	Nitrate as nitrogen source:	
2	$1/28\ NO_3^- + 5/28\ CO_2 + 29/28\ H^+ + e^- = 1/28\ C_5H_7O_2N + 11/28\ H_2O$	
	Reactions for electron acceptors (R_a)	
	Oxygen:	
3	$1/4\ O_2 + H^+ + e^- = 1/2\ H_2O$	−18.675
	Nitrate:	
4	$1/5\ NO_3^- + 6/5\ H^+ + e^- = 1/10\ N_2 + 3/5\ H_2O$	−17.128
	Sulfate:	
5	$1/8\ SO_4^{2-} + 19/16\ H^+ + e^- = 1/16\ H_2S + 1/16\ HS^- + 1/2\ H_2O$	5.085
	Carbon dioxide (methane fermentation):	
6	$1/8\ CO_2 + H^+ + e^- = 1/8\ CH_4 + 1/4\ H_2O$	5.763
	Reactions for electron donors (R_d)	
	Organic donors (heterotrophic reactions)	
	Domestic wastewater:	
7	$9/50\ CO_2 + 1/50\ NH_4^+ + 1/50\ HCO_3^- + H^+ + e^- = 1/50\ C_{10}H_{19}O_3N + 9/25\ H_2O$	7.6
	Protein (amino acids, proteins, nitrogenous organics):	
8	$8/33\ CO_2 + 2/33\ NH_4^+ + 31/33\ H^+ + e^- = 1/66\ C_{16}H_{24}O_5N_4 + 27/66\ H_2O$	7.7

9 Carbohydrates (cellulose, starch, sugars):
$1/4\ CO_2 + H^+ + e^-$ $= 1/4\ CH_2O + 1/4\ H_2O$ 10.0

10 Grease (fats and oils):
$4/23\ CO_2 + H^+ + e^-$ $= 1/46\ C_8H_{16}O + 15/46\ H_2O$ 6.6

11 Acetate:
$1/8\ CO_2 + 1/8\ HCO_3^- + H^+ + e^-$ $= 1/8\ CH_3COO^- + 3/8\ H_2O$ 6.609

12 Propionate:
$1/7\ CO_2 + 1/14\ HCO_3^- + H^+ + e^-$ $= 1/14\ CH_3CH_2COO^- + 5/14\ H_2O$ 6.664

13 Benzoate:
$1/5\ CO_2 + 1/30\ HCO_3^- + H^+ + e^-$ $= 1/30\ C_6H_5COO^- + 13/20\ H_2O$ 6.892

14 Ethanol:
$1/6\ CO_2 + H^+ + e^-$ $= 1/12\ CH_3CH_2OH + 1/4\ H_2O$ 7.592

15 Lactate:
$1/6\ CO_2 + 1/12\ HCO_3^- + H^+ + e^-$ $= 1/12\ CH_3CHOHCOO^- + 1/3\ H_2O$ 7.873

16 Pyruvate:
$1/5\ CO_2 + 1/10\ HCO_3^- + H^+ + e^-$ $= 1/10\ CH_3COCOO^- + 2/5\ H_2O$ 8.545

17 Methanol:
$1/6\ CO_2 + H^+ + e^-$ $= 1/6\ CH_3OH + 1/6\ H_2O$ 8.965

Inorganic donors (autotrophic reactions)

18 $Fe^{3+} + e^-$ $= Fe^{2+}$ -17.780

19 $1/2\ NO_3^- + H^+ + e^-$ $= 1/2\ NO_2^- + 1/2\ H_2O$ -9.43

20 $1/8\ NO_3^- + 5/4\ H^+ + e^-$ $= 1/8\ NH_4^+ + 3/8\ H_2O$ -8.245

21 $1/6\ NO_2^- + 4/3\ H^+ + e^-$ $= 1/6\ NH_4^+ + 1/3\ H_2O$ -7.852

22 $1/6\ SO_4^{2-} + 4/3\ H^+ + e^-$ $= 1/6\ S + 2/3\ H_2O$ 4.657

23 $1/8\ SO_4^{2-} + 19/16\ H^+ + e^-$ $= 1/16\ H_2S + 1/16\ HS^- + 1/2\ H_2O$ 5.085

Table 2.1, continued

Reaction Number	Half Reaction	$\Delta G°$ (w)[a] kcal per Electron Equivalent
24	$1/4\ SO_4^{2-} + 5/4\ H^+ + e^- = 1/8\ S_2O2/3^- + 5/8\ H_2O$	5.091
25	$H^+ + e^- = 1/2\ H_2$	9.670
26	$1/2\ SO_4^{2-} + H^+ + e^- = 1/2\ SO2/3^- + 1/2\ H_2O$	10.595

[a] Reactants and products at unit activity except $(H^+) = 10^{-7}$.

Adding the half reactions yields:

$$(1/50)C_{10}H_{19}O_3N + (9/25)H_2O + (1/4)O_2 + H^+ + e^- =$$
$$(9/50)CO_2 + (1/50)NH_4^+ + (1/50)HCO_3^- + H^+ + e^- + (1/2)H_2O \quad (2.4)$$

Now, multiplying Equation 2.4 by 50 yields:

$$C_{10}H_{19}O_3N + 18 H_2O + 12.5 O_2 = 9 CO_2 + NH_4^+ + HCO_3 \quad (2.5)$$

It may be seen from Equations 2.4 and 2.5 that each mole of wastewater utilized results in the transfer of 50 electrons. The energy yielded by the reactions is:

$$\Delta G^\circ = \Delta G^\circ \text{ of products} - \Delta G^\circ \text{ of reactants} \quad (2.6)$$

Thus, the energy available from the aerobic metabolism of wastewater is:

$$\Delta G^\circ = \frac{50 \text{ electrons}}{\text{mol wastewater}} \times \left(-18.675 \frac{\text{kcal}}{e^-} - 7.6 \frac{\text{kcal}}{e^-} \right) \quad (2.7)$$

$$\Delta G^\circ = -1313.75 \text{ kcal/mol wastewater}$$

In a similar manner to Example 2.1 we can show the energy yield from wastewater under anoxic conditions (NO_3^- as electron acceptor) to be -1236.4 kcal/mol wastewater. Under anaerobic conditions (CO_2 as electron acceptor) the available energy can be shown to be -91.85 kcal/mol wastewater. While the energies available from aerobic and anoxic metabolisms are comparable, the energy available from anaerobic metabolism is an order of magnitude less.

Synthesis

The waste or "food" is used for both energy and synthesis. To determine the amount of electron acceptor (oxygen) required and the end products produced, the engineer needs to know what portion of the food is used for energy. To determine nutrient requirements (N,P) and sludge production he must know what portion of the food will be converted to cell mass (synthesis). McCarty [4] has approached this problem by writing a balanced half equation in the form:

$$R = f_sR_c + f_eR_e - R_d \quad (2.8)$$

where   R   =   overall reaction

          R_c =   half reaction for synthesis of bacterial cells
                 (assumed to be $C_5H_7O_2N$)

          R_e =   half reaction for electron acceptor

          R_d =   half reaction for electron donor

          f_s =   fraction of electron donor used for synthesis

          f_e =   fraction of electron donor used for energy

$$f_s + f_e = 1 \qquad\qquad (2.9)$$

Table 2.2 presents maximum values of f_s for various substrates. The values of $(f_s)_{max}$ given represent young, rapidly growing cultures. For very old cultures, f_s values may be as low as 20% of those shown. The above equations may be used to determine the mass balance for a biological system.

Example 2.2. Let us consider aerobic biological treatment of domestic wastewater, and utilize this technique to determine the oxygen required and the

Table 2.2. Typical Values for $(f_s)_{max}$ for Bacterial Reactions [5]

Electron Donor	Electron Acceptor	$(f_s)_{max}$
Heterotrophic reactions		
Carbohydrate	O_2	0.72
Carbohydrate	NO_3^-	0.60
Carbohydrate	SO_4^{2-}	0.30
Carbohydrate	CO_2	0.28
Protein	O_2	0.64
Protein	CO_2	0.08
Fatty Acid	O_2	0.59
Fatty Acid	SO_4^{2-}	0.06
Fatty Acid	CO_2	0.05
Methanol	NO_3^-	0.36
Methanol	CO_2	0.15
Autotrophic reactions		
S	O_2	0.21
$S_2O_3^{2-}$	O_2	0.21
$S_2O_3^{2-}$	NO_3^-	0.20
NH_4^+	O_2	0.10
H_2	O_2	0.24
H_2	CO_2	0.04
Fe^{2+}	O_2	0.07

sludge produced. Assume that the nitrogen source for synthesis is ammonia and that extended aeration, typical of oxidation ditch design, is used. Since extended aeration results in an "old" culture, f_s will be taken as 20% of the $(f_s)_{max}$ value from Table 2.2. Since no value of $(f_s)_{max}$ is given for wastewater, we must develop a value based on a typical composition of domestic wastewater. Domestic wastewater may be assumed to be approximately 50% protein, 40% carbohydrate and 10% fat [6,7]. Referring to Table 2.2 we can develop a value of $(f_s)_{max}$:

$$(f_s)_{max} = \% \text{ protein} \times (f_s)_{max, \text{ protein}} + \% \text{ carbohydrate} \times$$
$$(f_s)_{max, \text{ carbohydrate}} + \% \text{ fat} \times (f_s)_{max, \text{ fat}} \tag{2.10}$$

$$(f_s)_{max} = (0.50 \times 0.64) + (0.40 \times 0.72) + (0.59 \times 0.10)$$

$$(f_s)_{max} = 0.67$$

In our example, f_s is taken to be 20% of $(f_s)_{max}$ or 0.13. From Equation 2.9, f_e may then be seen to be 0.87.

The appropriate half reactions are now selected from Table 2.1:

$$R_c: \quad (1/5)CO_2 + (1/20)HCO_3^- + (1/20)NH_4^+ + H^+ + e^- =$$
$$(1/20)C_5H_7O_2N + (9/20)H_2O \tag{2.11}$$

$$R_e: \quad (1/4)O_2 + H^+ + e^- = (1/2)H_2O \tag{2.12}$$

$$R_d: \quad (9/50)CO_2 + (1/50)NH_4^+ + (1/50)HCO_3^- + H^+ + e^- =$$
$$(1/50)C_{10}H_{19}O_3N + (9/25)H_2O \tag{2.13}$$

In accordance with Equation 2.8, Equations 2.11, 2.12 and 2.13 are combined and simplified, yielding:

$$C_{10}H_{19}O_3N + 10.875\ O_2 = 0.325\ C_5H_7O_2N + 0.675\ NH_4^+ +$$
$$+ 0.675\ HCO_3^- + 7.70\ CO_2 + 6.675\ H_2O \tag{2.14}$$

Thus, we can say that 1 mol of domestic sewage will yield 0.325 mol of microorganisms ($C_5H_7O_2N$) and will require 10.875 mol of O_2. These results will be much more useful in design and evaluation of oxidation ditches if we express the results in more common units.

A convenient method of expression might be mg oxygen required/mg COD and mg volatile suspended solids (VSS)/mg COD. First, let us consider oxygen. We require 10.875 mol O_2 per mol $C_{10}H_{19}O_3N$. The COD of wastewater may be calculated as:

$$COD = (10 + 19/4 - 3/2) = 13.25 \; \frac{\text{mol } O_2}{\text{mol } C_{10}H_{19}O_3N} \tag{2.15}$$

$$COD = \frac{13.25 \text{ mol } O_2}{\text{mol } C_{10}H_{19}O_3N} \; x \; \frac{32 \text{ g}}{\text{mol } O_2} = \frac{424 \text{ g COD}}{\text{mol } C_{19}H_{19}O_3N} \tag{2.16}$$

Thus:

$$O_2 \text{ required} = 10.875 \; \frac{\text{mol } O_2}{\text{mol } C_{10}H_{19}O_2N} \; x \; \frac{1 \text{ mol } C_{10}H_{19}O_3N}{424 \text{ g COD}} \; x$$

$$\frac{32 \text{ g}}{\text{mol } O_2} \tag{2.17}$$

Therefore:

$$O_2 \text{ required} = 0.82 \text{ g } O_2/\text{g COD}$$

In a similar manner:

$$VSS_{produced} = 0.375 \; \frac{\text{mol } C_5H_7O_2N}{\text{mol } C_{10}H_{19}O_3N} \; x \; \frac{113 \text{ g } C_5H_7O_2N}{\text{mol } C_5H_7O_2N} \; x$$

$$\frac{\text{mol } C_{10}H_{19}O_3N}{424 \text{ g COD}} \tag{2.18}$$

Therefore:

$$VSS_{produced} = 0.10 \text{ g VSS/g COD}$$

The same approach may be used to determine nutrient requirements, alkalinity produced or destroyed, and changes in hydrogen ion concentration (pH).

BACTERIAL KINETICS

To properly design and operate biological treatment systems, including oxidation ditches, it is necessary to understand bacterial kinetics. The degradation of organics contained in the wastewater, as well as nitrification and denitrification, depends on the rate of bacterial growth.

The simplest system which may be considered is a batch system containing a pure culture of microorganisms. For pure culture microbial systems, growth may be represented in batch systems by a plot of the log of number of organisms vs time (Figure 2.1).

The same information may be more usefully expressed in terms of mass rather than number of organisms, as shown in Figure 2.2.

The lag and log growth phases are as described in Figure 2.1. In biological treatment systems the declining growth phase is assumed to be substrate-limited. The endogenous phase is substrate-limited and organisms metabolize protoplasm without replacement.

Batch systems pass through all phases shown in Figures 2.1 and 2.2. Continuous loop reactors are generally not batch but rather continuous flow

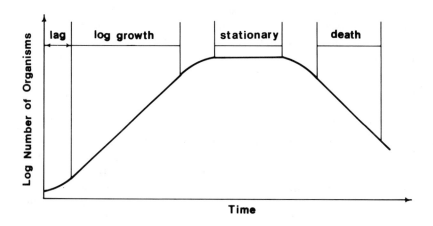

Figure 2.1. Growth in batch systems. Lag phase: this represents an acclimation of the organisms to the substrate. This phase need not always be present. Log growth phase: the growth rate is limited only by the microorganism's generation rate and its ability to process food. Stationary growth phase: reproduction rates are balanced by death rates. Growth may be limited by availability of substrate, accumulation of toxics, nutrients, or environmental factors such as O_2 or pH. Log death: substrate is exhausted, toxics accumulate and/or environmental conditions become unfavorable.

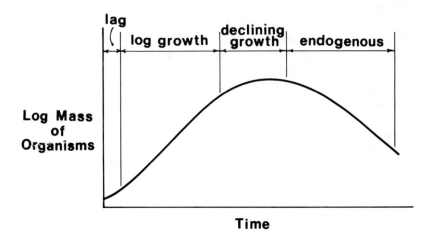

Figure 2.2. Growth in batch systems (expressed in terms of mass).

systems. Control of biological treatment systems is accomplished through control of substrate available to the microorganisms. To accomplish control through availability of substrate, it is necessary that all environmental conditions required for growth of microorganisms be maintained.

Environmental Conditions

The environmental conditions of concern include temperature, pH, dissolved oxygen and nutrients. Of these, the environmental engineer can control all but temperature. As will be seen, temperature effects may be accounted for in the design and operation of continuous loop reactors. However, it is usually impractical to attempt to control wastewater temperature. The dissolved oxygen level and the pH in the reactor may be controlled to any desired level. The major nutrients required for microbial growth are nitrogen and phosphorus. Based on the composition of microbial cells grown in the presence of adequate nutrients [4], the nitrogen requirement is approximately 1 mg/L per 20 mg/L BOD_5 [8]. The phosphorus requirement is approximately 1 mg/L PO_4-P per 100 mg/L BOD_5 [8]. Trace inorganic ions are also required [9]. It should be noted that the nitrogen and phosphrous requirements given are approximate. The actual nutrients required will depend on the amount of cell synthesis, which will in turn depend on process design.

Most municipal and domestic wastewaters contain the required nutrients in excess. In the case of industrial wastes, nitrogen and/or phosphorus may have to be added to the wastewater to satisfy nutrient requirements.

Assuming that all environmental conditions are satisfactorily maintained, it is then possible to maintain the biological system at any desired point on the batch system growth curve (Figures 2.1 and 2.2).

Kinetic Equations

From Figure 2.2, it may be seen that during the log growth phase, the rate of microorganism growth may be expressed as:

$$\frac{dX}{dt} = \mu X \tag{2.19}$$

where X = concentration of microorganisms, mg/L
μ = specific growth rate, day^{-1}
t = time, days

Continuous systems such as oxidation ditches are rarely maintained in the log growth phase. Rather, they are maintained at various points on the growth curve (Figure 2.2) depending on process modification. For example, oxidation ditches are usually designed to operate in the extended aeration mode, which is in the endogenous range. In order to determine the proper control of the bioreactor, it is necessary to develop relationships between growth rates and substrate utilization.

The most common model used to relate microbial growth with substrate utilization is that of Monod [10-12]. Monod found that in pure cultures and continuous systems, the relationship between bacterial growth and substrate availability, in systems with a single growth-limiting substrate, could be expressed empirically as:

$$\mu = \mu_m \cdot \frac{S}{K_s + S} \tag{2.20}$$

where μ = specific growth rate, day^{-1}
μ_m = maximum specific growth rate, day^{-1}
S = substrate concentration, mg/L
K_s = half-velocity constant, mg/L (The concentration of limiting substrate at which the specific growth rate equals one half the maximum specific growth rate.)

As shown in Figure 2.3, the microbial growth rate increases as the availability of substrate increases until a maximum specific growth rate (μ_m) is reached. At that point a factor other than substrate, such as generation rate or a nutrient, becomes growth-limiting.

The Monod model has been utilized to develop kinetic models of biological systems. Here, the Monod model will be used to develop kinetic models for the design and operation of oxidation ditches. It should be noted that a number of other models have been proposed [13-18]. Most of these models yield similar results when steady-state conditions exist. For highly variable influents, other models may be used which predict effluent substrate concentration as a function of influent concentration [14-18]. The Monod model as used by Lawrence and McCarty [19] has yielded acceptable results for wastewater systems and will be used in the following discussion.

We have already noted that:

$$\frac{dX}{dt} = \mu X \qquad (2.19)$$

Thus, combining Equation 2.19 and Equation 2.20 yields:

$$\left(\frac{dX}{dt}\right)_{growth} = \mu_m \frac{SX}{K_s + S} \qquad (2.21)$$

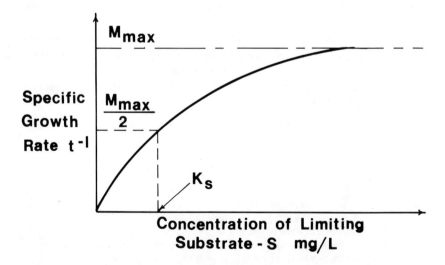

Figure 2.3. Effect of limiting substrate on microbial growth.

All of the available substrate is not used for synthesis; it has been shown [19] that for the log growth phase:

$$\left(\frac{dX}{dt}\right)_{growth} = -Y\left(\frac{dS}{dt}\right)_{bio.} \tag{2.22}$$

where   Y   =   maximum yield coefficient and (dX/dt) represents the mass of cells formed per mass of substrate utilized

$\left(\dfrac{dS}{dt}\right)_{bio}$ =   rate of substrate utilization due to biological activity

By combining the above equations and defining:

$$k = \frac{\mu_m}{Y} \tag{2.23}$$

We can express the change in substrate concentration due to biological activity as:

$$\left(\frac{dS}{dt}\right)_{bio} = \frac{kXS}{K_s + S} \tag{2.24}$$

In continuous loop reactors, and most other biological treatment systems, not all cells are in the log growth phase. We must reduce the growth rate described above to account for substrate utilized to provide the energy needed for cell maintenance as well as for death and predation by protozoans. While not strictly true [20–23], it is useful to assume that all reduction in the number of organisms is due to endogenous respiration and that the rate of endogenous decay is first-order and depends only on the mass of organisms present. With these assumptions, the rate of endogenous decay (dX/dt) can be expressed as:

$$\left(\frac{dX}{dt}\right)_{decay} = -k_d X \tag{2.25}$$

where k_d = endogenous decay coefficient, day^{-1}

If we now combine the effect of endogenous respiration with the growth equation:

$$\text{new growth rate} = \text{log growth rate} - \text{endogenous decay rate} \tag{2.26}$$

We get:

$$\left(\frac{dX}{dt}\right)_{net} = \frac{\mu_m SX}{K_s + S} - k_d X \tag{2.27}$$

or, in terms of substrate utilization:

$$\left(\frac{dX}{dt}\right)_{net} = -Y\left(\frac{dS}{dt}\right)_{bio} - k_d X \tag{2.28}$$

The constants utilized in these expressions are temperature-dependent. The values of the constants increase with temperature. The effect of temperature on the overall process involved in continuous loop reactors is more complex. Temperature affects biochemical reaction rates, chemical reaction rates, solubility of gases, gas transfer rates and settleability of solids. The effect on the overall process of temperature may be estimated by [2]:

$$\frac{R_T}{R_{20}} = \Theta^{(T-20)} \tag{2.29}$$

where R_T = rate at temperature T, °C
R_{20} = rate at 20°C
Θ = constant, usually 1.02 to 1.04 for CLRs

Application to Continuous Loop Reactors

Continuous loop reactors are neither completely mixed reactors nor plug flow reactors. Flow enters the reactor and is mixed with a large volume of reactor contents. However, the flow generally is arranged to enter the CLR just downstream of the point of discharge, and it must travel one loop before any of the feed can be discharged. CLRs are similar to plug flow reactors in this respect. Since the overall hydraulic retention time in the CLR is large, usually 20–24 hr, and the time to complete one loop at normal channel velocities is small, 15 to 30 min, the kinetics of CLRs are generally treated as those for complete mix reactors. Any small error introduced by this assumption acts as a factor of safety.

For a completely mixed reactor (Figure 2.4) we can write:

$$\left(\frac{dX}{dt}\right)V = QX_o - QX + V\left(\frac{dX}{dt}\right)_{net} \qquad (2.30)$$

where Q = flowrate, MGD
  X_o = influent microorganism concentration, mg/L volatile suspended solids (VSS)
  V = reactor volume, MG
  X = concentration of microorganisms in the reactor, mg/L MLVSS

By substituting the terms developed in the previous equations and by assuming (1) steady state, and (2) X_o is small and may be neglected, we can write:

$$Q/V = \frac{\mu_m S}{K_s + S} - k_d \qquad (2.31)$$

Considering substrate in a similar manner yields, at steady state:

$$S_o - S = \Theta \frac{kSX}{K_s + S} \qquad (2.32)$$

where S_o = influent substrate concentration, mg/L
  S = effluent substrate concentration, mg/L

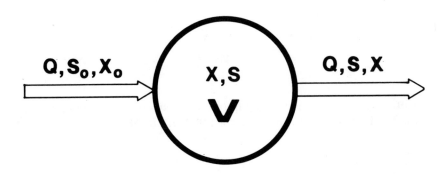

Figure 2.4. Completely mixed reactor.

Before considering the sludge recycle, which is used in any activated sludge process, including CLRs, it will be convenient to introduce some additional terms and to rearrange these equations. We will define:

$$\Theta_c = \frac{VX}{QX} \qquad (2.33)$$

where $\Theta_c = \dfrac{\text{mass of cells in reactor}}{\text{mass of cells wasted/unit time}}$, days

If we also define:

$$U = \frac{-(dS/dt)_{bio}}{X} \qquad (2.34)$$

where U = specific substrate utilization rate, mg/L/hr, or change in substrate concentration due to biological activity per unit mass of organism

We can show that:

$$\frac{1}{\Theta_c} = YU - k_d \qquad (2.35)$$

and

$$U = \frac{S_o - X}{\Theta_c X} \qquad (2.36)$$

We can define:

$$\Delta Y_{net} = \frac{Y}{1 + \Theta_c k_d} \qquad (2.37)$$

We often refer to organic loadings in terms of food-to-microorganism (F/M) ratio. The F/M ratio is usually expressed as lb BOD_5/lb MLVSS/d.

$$F/M = \frac{S_o}{\Theta X} \qquad (2.38)$$

If we define the overall process efficiency as:

$$E = \frac{S_o - S}{S_o} \times 100 \qquad (2.39)$$

where E = process efficiency, %

F/M, U and process efficiency are related by:

$$U = \frac{F/M \times E}{100} \qquad (2.40)$$

Two additional rearrangements are useful before considering solid recycle. We can show that:

$$X = \frac{Y(S_o - S)}{1 + k_d \, \Theta_c} \qquad (2.41)$$

and

$$S = \frac{K_s \, (1 + \Theta + k_d)}{\Theta_c (Yk - k_d) - 1} \qquad (2.42)$$

Thus, for completely mixed systems, if we know the kinetic coefficients $(Y, k, k_d$ and $K_s)$, we can predict X and S.

Sludge Recycle

Let us now consider sludge recycle while continuing to treat the CLR as a complete mix reactor from a kinetics standpoint. A schematic of the system is shown as Figure 2.5.

The following assumptions are made in the analysis:

1. The reactor is completely mixed.
2. The concentration of microorganisms in the influent (X_o) is small and may be assumed to be zero.
3. All waste stabilization occurs in the bioreactor.
4. The volume used for calculating Θ_c is the bioreactor volume only.

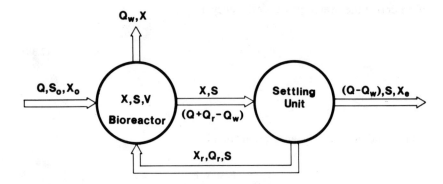

Figure 2.5. CLR schematic.

It also should be noted that we will take the influent substrate concentration to be the total BOD, since the solid material can be solubilized and utilized by the organisms. The effluent substrate concentration predicted by the equations is soluble BOD. Thus, in considering effluent quality the BOD of biomass contained in the effluent must be added.

If sludge is wasted directly from the bioreactor, as shown in Figure 2.5, then:

$$\Theta_c = \frac{VX}{Q_wX + (Q - Q_w) X_e} \qquad (2.43)$$

but since $X \ggg X_e$

$$\Theta_c \cong \frac{V}{Q_w} \qquad (2.44)$$

If sludge is wasted from the return sludge line:

$$\Theta_c = \frac{VX}{Q_wX_r + QX_e} \qquad (2.45)$$

and

$$\Theta_c \cong \frac{VX}{Q_wX_r} \qquad (2.46)$$

where X_R = concentration of microorganisms in return sludge, mg/L

By the same mass balance techniques as utilized before, we can show that:

$$\frac{1}{\Theta_c} = YU - k_d \tag{2.47}$$

and

$$X = \frac{\Theta_c}{\Theta} \frac{Y(S_o - S)}{1 + k_d\Theta_c} \tag{2.48}$$

Or in a more useful form:

$$XV = \frac{YQ(S_o - S)\,\Theta_c}{1 + k_d\Theta_c} \tag{2.49}$$

In this form it may be seen that for a given system the quantity XV is a constant but both X (MLVSS) and V (reactor volume) may be varied to optimize design.

Sludge Production

The biological sludge production in a CLR can be estimated from solids yield and endogenous decay. The biomass produced may be estimated from:

$$\left(\frac{dX}{dt}\right)_{growth} = Y(S_o - S) \times Q \tag{2.50}$$

and the biomass destroyed from:

$$\left(\frac{dX}{dt}\right)_{decay} = -k_d XV \tag{2.51}$$

and the net sludge produced:

$$\left(\frac{dX}{dt}\right)_{net} = Y(S_o - S) Q - k_d XV \tag{2.52}$$

For steady-state continuous loop reactors the waste activated sludge (P_X) is equal to dX/dt_{net}.

APPLICATION OF KINETICS
TO OXIDATION DITCHES

To apply the kinetics developed in this chapter to oxidation ditches, it is necessary to define the type of reactor that we call an oxidation ditch. In wastewater treatment two types of reactors are commonly considered: the completely mixed reactor and the plug flow reactor.

In a completely mixed reactor, it is assumed that the influent stream is instantaneously mixed with the contents of the reactor. Therefore, the effluent concentration of any parameter is equal to the concentration of that parameter in the reactor. In a plug flow reactor, it is assumed that no longitudinal dispersion occurs. Each particle of influent moves along the plug flow reactor without changing position relative to the other particles in the reactor. Thus, the concentration of any given parameter changes along the length of the reactor. It should be noted that a plug flow reactor is equivalent to an infinite series of completely mixed reactors. A much more detailed discussion of reactor types may be found in the literature [1,2,8,24-26].

The type of reactor chosen has an effect on the effluent quality and efficiency of a biological treatment system. A plug flow system can be shown to be more efficient in substrate removal than a completely mixed system [19,27,28]. This increase in efficiency may be seen in Figure 2.6 [19].

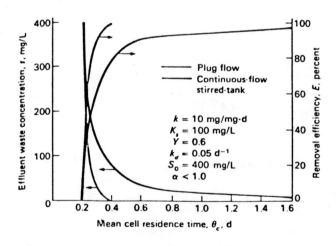

Figure 2.6. Comparison of completely mixed and plug flow reactors [19].

Completely mixed systems offer other advantages, however. The major advantage is that slug loads of toxic or inhibitory materials are rapidly mixed with the entire reactor contents and diluted, thus minimizing their effects. Since in practice neither instantaneous complete mixing nor perfect plug flow can be attained, the differences between the reactor types are minimized.

Continuous loop reactors combine features of both plug flow and completely mixed reactors. As in plug flow reactors, the incoming waste flow enters the reactor just downstream of the point of discharge. The flow then mixes with the entire reactor contents—a technique of completely mixed reactors. Characterization of the continuous loop reactor, however, is based on the short time necessary to complete one loop and the large overall hydraulic retention time. Because this is similar to completely mixed reactors [29,30], completely mixed reactors are commonly used to model continuous loop reactors.

Using the completely mixed reactor to describe an oxidation ditch, we can now develop the kinetic equations needed to describe the design and operation of oxidation ditches. Let us consider the continuous loop reactor shown in Figure 2.7.

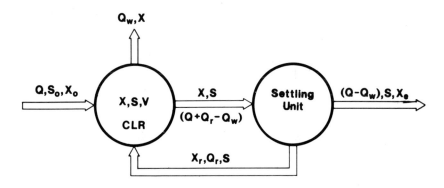

Figure 2.7. Oxidation ditch schematic. Q = influent flow rate, (MGD); S_o = total influent substrate concentration, (mg/L BOD_5); X_o = influent microorganism concentration, (mg/L VSS); V = reactor volume (million gal); S = reactor soluble substrate concentration, (mg/L BOD_5); Q_w = waste activated sludge (WAS) flow rate, (MGD); Q_R = return activated sludge (RAS) flow rate, (MGD); X_R = return activated sludge microorganism concentration, (mg/L VSS); and X_e = effluent microorganism concentration (mg/L VSS); X = reactor microorganism concentration (mg/L MLVSS).

In Figure 2.7, it has been assumed that sludge wasting is from the CLR. Sludge may also be wasted from the return activated sludge line. In the analysis of the CLR system shown in Figure 2.7, the following assumptions are made:

1. The reactor is completely mixed.
2. The influent microorganism concentration (X_0) is very small compared to the reactor microorganism concentration (X) and therefore may be assumed to be zero.
3. All utilization of substrate occurs in the CLR.
4. The volume used in calculations is the volume of the CLR.

The first two assumptions are close enough to the true situation to introduce little error. The last two assumptions are not strictly correct. Contact between living organisms and substrate occurs in the settling unit. Some substrate utilization therefore must occur in the clarifiers. The amount of BOD removal in the clarifier is small, however, due to lack of mixing and limited availability of oxygen. It is also clear that some volume of biomass is contained in the settling unit at any given time. The volume of biomass in the clarifier is small compared to the volume contained in the CLR in most systems. Thus, the error incurred by using this assumption is also small. The last two assumptions are conservative and will yield calculated BOD removals slightly lower than actual removals.

It should also be noted that in the development of kinetic equations the influent BOD_5 to the reactor is taken as total BOD_5. While the microorganisms cannot directly use organic material as food in solid form, extracellular enzymes will convert this material to soluble form. The equations developed will predict soluble BOD_5 in the effluent. Thus, in considering effluent quality the BOD of biomass contained in the effluent must be considered in order to derive an effluent BOD_5 concentration for design.

Example 2.3. The effluent requirement for a new oxidation ditch treatment system is 30 mg/L BOD_5 and 30 mg/L TSS. Assuming that the effluent suspended solids are 70% volatile, determine the value of S (effluent soluble BOD_5) to be used for design.

The effluent contains soluble BOD and BOD in the form of microorganisms. Taking an organism as $C_5H_7NO_2$, the BOD_L of the organism can be shown to be 1.42 g BOD_5/g VSS. This also assumes effluent VSS are microorganisms. This is a valid assumption.

$$BOD_L \text{ of VSS} = 0.70 \times 30 \times 1.42 = 29.8 \text{ mg/L}$$

Assuming a BOD rate constant of 0.20 day^{-1},

$$BOD_5 \text{ of VSS} = 29.8 \text{ mg/l} \times (1 - e^{-0.20 \times 5}) = 18.8 \text{ mg/L}$$

Therefore, the design soluble BOD_5 is:

$$S = 30 \text{ mg/L} - 18.8 \text{ mg/L} = 11.2 \text{ mg/L}$$

Before developing the equations for design and operation of oxidation ditches, it will be useful to define some additional terms. It has become increasingly common to express the design and operation of biological treatment systems in terms of mean cell residence time (Θ_c). Mean cell residence time may be defined as the mass of cells in the reactor divided by the mass of cells wasted from the reactor per unit time. Thus, with sludge being wasted as shown in Figure 2.5, Θ_c may be expressed as:

$$\Theta_c = \frac{VX}{Q_w X + (Q - Q_w) X_e} \tag{2.53}$$

where Θ_c = mean cell residence time, days

In real systems, the concentration of cells in the reactor (X) is much greater than the concentration of cells in the effluent (X_e). Also, the influent flow rate (Q) is much greater than the WAS flow rate (Q_w). Therefore, Θ_c may be approximated by:

$$\Theta_c \cong \frac{W}{Q_w} \tag{2.54}$$

If sludge is wasted from the return activated sludge line rather than directly from the reactor, we can express Θ_c as:

$$\Theta_c = \frac{VX}{Q_w X_R + Q X_e} \tag{2.55}$$

or

$$\Theta_c \cong \frac{VX}{Q_w X_R} \tag{2.56}$$

Returning to Figure 2.7, we can now write a mass balance around the oxidation ditch and its clarifier. Writing a mass balance in terms of microorganisms yields:

$$
\begin{array}{l}
\text{rate of increase} \\
\text{of microorganisms} \\
\text{in the system}
\end{array}
=
\begin{array}{l}
\text{rate of microorganisms} \\
\text{entering the system}
\end{array}
-
\begin{array}{l}
\text{rate of microorganisms} \\
\text{leaving the system}
\end{array}
$$

$$
+
\begin{array}{l}
\text{net growth rate of} \\
\text{microorganisms} \\
\text{within the system}
\end{array}
\tag{2.57}
$$

It has been shown [20,21,31] that the net yield coefficient (Y_{net}) may be expressed as:

$$
Y_{net} = \frac{Y}{1 + k_d \Theta_c}
\tag{2.58}
$$

Substituting into Equation 2.33 yields:

$$
V\left(\frac{dX}{dt}\right)_{bio} = QX_o - Q_w X - (Q - Q_w)X_e + V \frac{\left(\dfrac{dS}{dt}\right)_{bio} Y}{1 + k_d \Theta_c}
\tag{2.59}
$$

where $\left(\dfrac{dX}{dt}\right)_{bio}$ = increase in microorganisms per unit time as a result of biological activity

$\left(\dfrac{dS}{dt}\right)_{bio}$ = rate of substrate utilization due to biological activity

Assuming that (1) $X_o \cong 0$, (2) $(Q - Q_w) X_e \cong 0$, and (3) steady-state conditions exist [that is $(dX/dt)_{bio}$ is equal to zero], the expression may be written as:

$$
X = \left(\frac{\Theta_c}{\Theta}\right) \frac{Y (S_o - S)}{1 + k_d \Theta_c}
\tag{2.60}
$$

where Θ = hydraulic retention time, days

Noting that the hydraulic retention time is equal to the reactor volume divided by the influent flow rate, we can rewrite Equation 2.60 in a more useful form:

$$XV = \frac{YQ\ (S_o - S)\ \Theta_c}{1 + k_d \Theta_c} \tag{2.61}$$

Equation 2.61 may be conveniently used in the design and operation of oxidation ditches for carbonaceous BOD removal.

Example 2.4. It is desired to size a CLR to treat 1 MGD of wastewater having an influent BOD_5 of 200 mg/L. The required effluent soluble BOD_5 had been determined to be 10 mg/L by the methods shown in Example 2.3. Kinetic studies have yielded values of $Y = 0.60$ mg VSS/mg BOD_5 and k_d = 0.05 day^{-1}. For the required level of treatment a value of Θ_c from 5-7 days is commonly used. We will use $\Theta_c = 6$ days.

$$XV = \frac{YQ(S_o - S)\ \Theta_c}{1 + k_d \Theta_c} \tag{2.62}$$

Substituting:

$$XV = \frac{0.6\ \dfrac{\text{mg VSS}}{\text{mg BOD}_5}\ \times\ 1\ \text{MGD}\ \times\ (200\text{--}10)\ \text{mg/L}\ \times\ 6\ \text{days}}{1\ +\ 0.05\ \text{day}^{-1}\ \times\ 6\ \text{days}}$$

$$XV = 526.2\ \text{mg/L} \times \text{MG}$$

We are now free to select any combination of X and V which when multiplied will yield 526.2 mg/L x MG. Selecting a design value of MLSS of 3000 mg/L, of which 70% are assumed to be volatile:

$$MLVSS = X = 0.7\ \times\ 3000\ \text{mg/L} = 2100\ \text{mg/L}$$

Then:

$$V = \frac{526.2\ \text{mg/L}\ \times\ \text{MG}}{2100\ \text{mg/L}} = 0.25\ \text{MG}$$

Thus the required volume of the oxidation ditch is 250,000 gal. It should be noted that in this example a hydraulic retention time of 6 hr was calculated. It should be noted that use of nitrification/denitrification, as well as considerations of sludge production, operational stability and safety factors, will result in the longer hydraulic retention times typical of extended aeration systems.

KINETIC DATA

The successful design of any biological treatment system depends on the use of appropriate kinetic data. For domestic wastewaters, typical kinetic data are often found in the literature. Even when considering domestic wastewaters, these data must be used in conjunction with considerable judgment on the part of the design engineer. There is no such thing as a "typical" wastewater.

Whenever possible, it is highly desirable to determine the kinetic constants by bench- or, preferably, pilot-scale investigations. Several methods for carrying out these investigations may be found in the literature [1,2,8,26,32,33] and are not repeated here. A method for kinetic parameter determination from thermodynamic considerations has been presented by McCarty [4].

When it is not possible to carry out bench- or pilot-scale investigations on the waste to be treated, due to unavailability of the waste, financial constraints, etc., the designer is forced to rely on the literature. Table 2.3 presents typical data from the literature.

Table 2.3. Kinetic Data for Biological Treatment

Substrate	Y_1 mg VSS/ mg substrate	k_d, day^{-1}	K_s, mg/L	μ_{max}, day^{-1}	Basis	Source
Domestic Waste	0.5	0.055			BOD$_5$	21
Domestic Waste	0.67	0.048			BOD$_5$	34
Domestic Waste			60	9.6	COD	36
Domestic Waste	0.50	0.06	120	13.2	BOD$_5$	36
Domestic Waste	0.67	0.07	22	3.84	COD	36
Soft Drink Waste	0.35	0.031	0.31	0.35	COD	35
Skim Milk	0.48	0.045	100	2.38	BOD$_5$	19
Synthetic Waste	0.65	0.18			BOD$_5$	37
Pulp & Paper	0.47	0.20			BOD$_5$	36
Shrimp Processing	0.50	1.60	85.5		BOD$_5$	36

Table 2.4. Kinetic Constants for Municipal Wastewater at 20°C [38]

Parameter	Units	Values
Y	mg VSS/mg COD	0.35-0.45
k_d	day^{-1}	0.05-0.10
K_s	mg/L	25-100

It may be clearly seen from Table 2.3 that large variations occur in the values obtained for kinetic constants from various studies and wastewaters. Even the recommended kinetic constants for municipal wastewater [38] (Table 2.4) show wide variations.

In the design of oxidation ditches, the oxidation of ammonia and the biological removal of nitrogen are often of concern. In many cases the design of the continuous loop reactor will be controlled by nitrification/denitrification considerations. Biological nitrification and denitrification are discussed in Chapter 4.

REFERENCES

1. Benefield, L. D., and C. W. Randall, *Biological Process Design for Wastewater Treatment* (Englewood Cliffs, NJ: Prentice-Hall, Inc., 1980).
2. Metcalf and Eddy, *Wastewater Engineering*, 2nd Edition, (New York: McGraw-Hill Book Company, 1979).
3. Setlow, R. B., and C. F. Polland, *Molecular Biophysics* (Reading, PA, Addison-Wesley Publishing Co., Inc., 1964).
4. McCarty, P. L., "Stoichiometry of Biological Reactions," *Progress in Water Technology* (London: Pergamon Press, 1975), p. 7.
5. Sawyer, C. M., and P. L. McCarty, *Chemistry for Environmental Engineering*, 3rd Edition (New York: McGraw-Hill Book Company, 1978).
6. Painter, H., M. Viney and A. Bywaters, "Composition of Sewage and Sewage Effluents," *Journal and Proceedings, Institute of Sewage Purification*, 302 (1961).
7. Hunter, J. V., "The Organic Composition of Various Domestic Sewage Fractions," PhD Thesis, Rutgers, The State University (1961).
8. Sundstrom, D. W., and H. E. Klei, *Wastewater Treatment* (Englewood Cliffs, N.J.: Prentice-Hall, Inc., 1979).
9. Wood, D. K., and G. Tchobanoglous, "Trace Elements in Biological Waste Treatment," *J. Water Poll. Control Fed.*, 47: 7 (1975).
10. Monod, J., *Recherches sur la Croissance des Cultures Bacteriennes* (Paris: Herman et Cie., 1942).
11. Monod, J., "La Technique of Culture Continue; Theorie et Applications," *Annals Institue Pasteur*, 79:340 (1950).

12. Monod, J., "The Growth of Bacterial Cultures," *Ann. Rev. Microbiol.*, III, (1949).
13. Grady, C. P. N., Jr., and D. R. Williams, "Effects of Influent Substrate Concentration on the Kinetics of Natural Microbial Populations in Continuous Cultures," *Water Res.*, 9:171 (1975).
14. Grau, P., M. Dohanyos and J. Chudoba, "Kinetics of Multicomponent Substrate Removal by Activated Sludge," *Water Res.*, 9:637 (1975).
15. Benefield, L. D., and C. W. Randall, "Evaluation of a Comprehensive Kinetic Model for the Activated Sludge Process," *J. Water Poll. Control Fed.*, 49:1636 (1977).
16. Adams, C. D., W. W. Eckenfelder and J. Hovius, "A Kinetic Model for Design of Completely Mixed Activated Sludge Treating Variable Strength Industrial Wastewaters," *Water Res.*, 9(1):37 (1975).
17. Elmaleh, S., and R. Ben Aim, "Influence Sur la Cinétique Biochimique de la Concentration en Carbone Organique a l'entree d'un Réacteur Developpant une Polyculture Microbienne en Mélange Parfait," *Water Res.*, 10:1005 (1976).
18. Vandevenne, L., and W. W. Eckenfelder, Jr., "A Comparison of Models for Completely Mixed Activated Sludge Treatment Design and Operation," *Water Res.*, 14:561 (1980).
19. Lawrence, A. W., and P. L. McCarty, "A Unified Basis for Biological Treatment Design and Operation," *J. San. Eng. Div., ASCE,* 96(SA3): 757 (1970).
20. Weston, R. F., and W. W. Eckenfelder, Jr., "Application of Biological Treatment to Industrial Wastes, I. Kinetics and Equilibria of Oxidative Treatment," *Sew. Ind. Wastes,* 27:802 (1955).
21. Heukelekian, H., H. E. Orford and R. Manganelli, "Factors Affecting the Quality of Sludge Production in the Activated Sludge Process," *Sew. Ind. Waste*, 23:8 (1951).
22. Dawes, E. A., and D. W. Ribbons, "The Endogenous Metabolism of Microorganisms," *Ann Rev. Microbiol.*, 16:241 (1962).
23. Mallette, F. M., "Validity of the Concept of Energy Maintenance," *Ann. N.Y. Acad. Sci.*, 102:521 (1963).
24. Denigh, K. A., and J. C. R. Turner, *Chemical Reactor Theory*, 2nd Edition (New York: Cambridge University Press, 1965).
25. Schroeder, E. D., *Water and Wastewater Treatment* (New York: McGraw-Hill Book Company, 1977).
26. Eckenfelder, W. W., Jr., *Principles of Water Quality Management* (Boston: C.B.I., 1980).
27. Grieves, R. B., W. P. Milbury and W. O. Pipes, "A Mixing Model for Activated Sludge," *J. Water Poll. Control Fed.*, 36(5):619-635 (1964).
28. Murphy, K. L., and P. L. Timpany, "Design and Analysis of Mixing for an Aeration Tank," *J. San. Eng. Div., ASCE,* 93(SA5):1-15 (1967).
29. Jacobs, A., "A New Loop in Aeration Tank Design," presented at Spring Meeting (July, 1974).
30. Stensel, H. D., and J. H. Scott, "Cost Effective Advanced Biological Treatment Systems," presented at Annual Meeting Nevada Water Pollution Control Association (Dec., 1977).

31. van Uden, N. "Transport-Limited Growth in the Chemostat and Its Competitive Inhibition; A Theoretical Treatment," *Archiv. Mikrobiol.* 58:145-154 (1967).

32. Irvine, R. L., and D. J. Schaezler, "Kinetic Analysis of Data from Biological Systems," *J. San. Eng. Div., ASCE,* 97(SA4):409-423 (1971).

33. Giona, A. R., et al., "Kinetic Parameters for Municipal Wastewater," *J. Water Poll. Control Fed.,* 51(5):999-1008 (1979).

34. Middlebrooks, E. J., and C. F. Garland, "Kinetics of Model and Field Extended-Aeration Tanks," *J. Water Poll. Control Fed.,* 40(4):586-612 (1968).

35. Yang, P. Y., and Y. K. Chen, "Operational Characteristics and Biological Kinetic Constants of Extended Aeration Process," *J. Water Poll. Control Fed.,* 49(4):678-688 (1977).

36. Mynhier, M. D., and C. P. L. Grady, Jr., "Design Graphs for Activated Sludge Process," *J. Environ. Eng. Div., ASCE,* 101:829 (1975).

37. McCarty, P. L., and C. F. Broderson, "Theory of Extended Aeration Activated Sludge," *J. Water Poll. Control Fed.,* 34(11):1095-1103 (1962).

38. Lawrence, A. W., "Modeling and Simulation of Slurry Biological Reactors," in *Mathematical Modeling for Water Pollution Control,* T. Keinath and M. Wanielista, Eds. (Ann Arbor, MI: Ann Arbor Science Publishers, Inc., 1975).

CHAPTER 3

PROCESS MODIFICATIONS

Though the great majority of continuous loop reactors (CLR) have employed the extended aeration mode of activated sludge treatment, there is no reason why a CLR cannot be used for almost any suspended growth biological system. Table 3.1 summarizes parameters for various activated sludge systems. CLRs have been used for conventional-rate systems and aerobic digestion. High-rate systems have been used for industrial wastes and in Europe. Multistage CLR systems have also appeared in Europe and Africa and, more recently, in North America.

The kinetic equations presented in Chapter 2 allow the design of a CLR for any rate activated sludge system or modification thereto. Nitrification and denitrification are discussed in Chapter 4.

A conventional activated sludge CLR might be used for large flows where the area requirements for extended aeration size basins are prohibitive. The same mixing characteristics persist in a conventional-rate CLR as in an extended-aeration–rate CLR. Power density and channel velocity are obviously increased as a result of a smaller channel per unit of organic loading. If channel velocity becomes excessive, rotor or turbine capacity will decrease as a result of the decreased relative velocity between rotor or turbine blades and the CLR contents. Primary clarification would normally precede conventional-rate CLRs.

One high-rate CLR in Austria is designed to be totally energy self-sufficient. Primary sludge is combined with waste activated sludge from the high–sludge-yield secondary process for anaerobic digestion. Digestor gas powers gas drives on air compressors and motor generators supplying recirculation pumps. Jet aerators are used for propulsion and mixing. Heat pumps using secondary clarifier water as the constant temperature medium augment waste heat from the gas motors to heat the anaerobic digesters.

Table 3.1 Common Design Parameters and Operating Characteristics of Single-Stage Activated Sludge Systems [1]

Process Type	Loading F/M, lb BOD$_5$/lb MLSS x day	SRT, days	lb BOD$_5$/1000 ft^3 x day[a] @ 3000 mg/l MLSS	BOD$_5$ Removal, %	Aerator Detention Time, hr	Nitrification Occurs	O$_2$ Required,[b] lb/lb BOD$_5$ removed	Recirculated Solids Rate, % Q	MLSS,[c] mg/L	O$_2$ Uptake, mg/g x hr MLSS	Waste Sludge, lb/lb BOD$_5$ removed
Extended aeration	≤0.05	≥30	10-15	90+	16-24	yes	1.4-1.6[d]	100-300	2,000-6,000	3-8	0.15-0.3
Conventional	0.15-0.4	4-8	20-60	90-95	4-8	possible	0.8-1.1[d]	30-100	1,500-4,000	7-15	0.4-0.6
High-rate	0.4-1.0	2-4	70-180	85-90	2-4	no	0.7-0.9	30-100	3,000-5,000	15-25	0.5-0.7
Modified aeration	1.5-3.0	<1	90-180+[e]	60-75	0.5-2	no	0.4-0.6	10-30	500-1,500	20-40	0.8-1.2
Contact-stabilization	0.15-0.5	3-10	30-70	85-95	1.0-3.0	possible	0.8-1.1	25-75	2,000-4,000	20-30	0.4-0.6
contact stabilization	0.5-2.0		90-180	85-95	3.0-6.0	no	0.4-0.6	50-100	6,000-10,000	10-30	
Single-stage nitrification	0.05-0.15	10-15	10-30	95+	6-12	yes	1.1-1.5	30-100	3,000-6,000	3-8	0.15-0.3

[a] Note: lb/1000 ft^3 x 4.883 = g/m^2.
[b] Density of O$_2$ @ 0°C and 760 mm = 0.089 lb/ft^3 (1.429 g/l).
[c] MLSS x 0.8 ≅ MLVSS.
[d] Additional oxygen must be added if nitrification takes lace.
[e] MLSS = 1000 mg/L.

PROCESS MODIFICATIONS
FOR BIOLOGICAL NUTRIENT REMOVAL

Biological nitrogen removal in CLRs is discussed in Chapters 2 and 7. Biological phosphorus removal may also be possible.

Barnard [2] in 1972 demonstrated nitrogen and phosphorus removal in a four-stage biological system as shown in Figure 3.1. In this system nitrified mixed liquor is recycled from the second basin to the first basin, where anaerobic or anoxic conditions promote nitrate removal from the recycled stream using raw sewage or primary effluent as the carbon source. To reduce the nitrate to lower levels, the third basin is maintained under anoxic conditions to promote further denitrification by endogenous sludge. Reaeration occurs in the final basin prior to clarification. Barnard found that phosphorus removal could be achieved in this process without chemical additives. He attributed the phosphorus removal to biological phosphorus uptake after sufficiently anaerobic conditions cause a release of phosphorus. The phosphorus is then taken up in the aerobic zones in excess amounts and incorporated in the biological cell mass, resulting in low soluble-phosphorus levels. If the phosphorus-rich sludge is wasted from the system and disposed of, high phosphorus removal is possible. An adaptation of this concept is shown in Figure 3.2.

Great care should be exercised in application of biological phosphorus removal schemes to various wastewaters. Barnard's work was done in South Africa at relatively low concentrations of phosphorus in the wastewater. Whenever possible, pilot studies are urged as a prudent course of action.

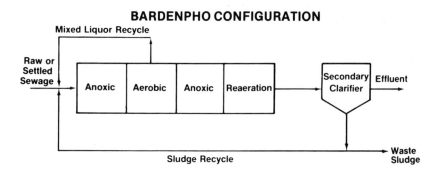

Figure 3.1 Biological phosphorus removal.

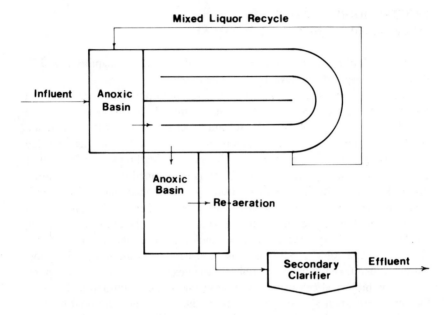

Figure 3.2 Biological phosphorus removal in a CLR.

REFERENCES

1. "Operation of Wastewater Treatment Plants," Water Pollution Control Federation, MOP 11, Washington, DC (1976).
2. Barnard, J. L. "Biological Nutrient Removal Without the Addition of Chemicals," *Water Res.* 9(516) (May/June, 1975).

CHAPTER 4

NITRIFICATION AND DENITRIFICATION

Nitrogen, usually in the form of ammonia and organic nitrogen, is contained in most wastewaters. As discussed in Chapter 2, nitrogen is present in domestic wastewaters far in excess of the quantities required for biological treatment. The presence of nitrogen and the form of nitrogen in treated wastewaters may be of concern for several reasons:

1. the effect of ammonia on disinfection
2. ammonia toxicity to aquatic organisms
3. the oxygen demand of unoxidized forms of nitrogen
4. the role of nitrogen in eutrophication
5. public health aspects of nitrate in water supplies

Each of these concerns is discussed briefly below.

PRESENCE OF NITROGEN AND AMMONIA

Effect of Ammonia on Disinfection

Chlorine is commonly used for disinfection of wastewater effluents. When chlorine is added to water containing ammonia, chloramines are formed, as shown in the following equations [1]:

$$NH_4^+ + HOCl \rightleftharpoons NH_2Cl + H_2O + H^+ \tag{4.1}$$

$$NH_2Cl + HOCl \rightleftharpoons NHCl_2 + H_2O \tag{4.2}$$

$$NHCl_2 + HOCl \longrightarrow NCl_3 + H_2O \tag{4.3}$$

Since combined chlorine is less effective than free chlorine (HOCl or OCl⁻) as a disinfectant, the presence of ammonia in a wastewater effluent results in more chlorine, or necessitates a longer detention time in the chlorine contact chamber, to achieve the same level of disinfection than would be required if ammonia were not present.

Aquatic Toxicity

Free ammonia (NH_3) is toxic to marine organisms at low concentrations. Ammonia in water is in equilibrium with the ammonium ion, as shown in Equation 4.4.

$$NH_4^+ + OH^- \rightleftharpoons NH_3 + H_2O \qquad (4.4)$$

As may be seen from Equation 4.4 the percent of ammonia present is a function of pH. A small change in pH results in a large change in the amount of unionized ammonia present. The U.S. Environmental Protection Agency (EPA) [2] has set the water quality criterion for the protection of freshwater aquatic life at 0.02 mg/L NH_3-N of unionized ammonia. The concentrations of total ammonia which will result in 0.02 mg/L of unionized ammonia as a function of pH are presented as Table 4.1. Since natural waters are often found in the pH range from 7.0 to 8.0, control of ammonia discharge may be required to avoid ammonia toxicity to aquatic organisms.

Nitrogen Oxygen Demand

When ammonia is discharged to receiving waters, it is converted by autotrophic organisms to nitrite and then nitrate. It has been estimated [3] that approximately 4.6 mg/L of oxygen is required for each mg/L of NH_3-N

Table 4.1. Concentration of Total NH_3-N Which Contains
0.02 mg/L of Unionized NH_3-N at 20°C [2]

pH	Total NH_3-N, mg/L
6.0	50.0
6.5	16.0
7.0	5.1
7.5	1.6
8.0	0.52
8.5	0.18

oxidized. Since a typical secondary effluent may contain approximately 25 mg/L of NH_3-N, the resulting nitrogenous oxygen demand of 115 mg/L will be far in excess of the approximately 45 to 60 mg/L of ultimate carbonaceous oxygen demand contained in such effluents. This high nitrogenous oxygen demand is the reason that many plants are required to provide nitrification of effluents prior to discharge. Nitrification will be defined as the conversion of ammonia to nitrate.

Eutrophication

Nitrogen is an essential nutrient for the growth of plants. Enrichment of a receiving water with nitrogen, in the presence of the other nutrients required for plant growth, may result in eutrophication. Eutrophication may be looked upon as an acceleration of the natural aging process of a lake.

Nitrogen may reach receiving waters from agricultural runoff, wastewater effluents and a variety of other sources. Control of nitrogen input from wastewater effluents requires removal of the nitrogen in all its forms. In continuous looped reactors this can be accomplished by biological nitrification and denitrification.

Public Health Effects

Nitrogen, in nitrate form, has been recognized since 1945 as the cause of a temporary blood disorder affecting infants called methemoglobinemia [4]. This disease can be fatal in small infants [5]. Nitrate, present in drinking water, can be converted to nitrite in the gastrointestinal tract of infants. The nitrite then combines with the hemoglobin in the blood and interferes with the ability to carry and transfer oxygen. For this reason, the primary drinking water standard for NO_3-N is 10 mg/L [6]. Nitrogen removal may be required for effluents which will reach potable water supplies.

BIOCHEMICAL REACTIONS IN NITRIFICATION

In aqueous systems ammonia is converted to nitrite by autotrophic bacteria, *Nitrosomonas*, and then to nitrate by *Nitrobacter*, which are also autotrophic. The oxidation reactions may be represented by [3]:

$$NH_4^+ + 1.5\,O_2 \xrightarrow{\quad Nitrosomonas \quad} 2\,H^+ + H_2O + NO_2^- \qquad (4.5)$$

and

$$\text{NO}_2^- + 0.5\,\text{O}_2 \xrightarrow{\quad\textit{Nitrobacter}\quad} \text{NO}_3^- \qquad (4.6)$$

Equations 4.5 and 4.6 may be added to yield an overall oxidation equation:

$$\text{NH}_4^+ + 2\,\text{O}_2 \longrightarrow \text{NO}_3^- + 2\,\text{H}^+ + \text{H}_2\text{O} \qquad (4.7)$$

It may be noted in Equation 4.7 that the oxidation of ammonia to nitrate results in the release of hydrogen ions and therefore a lowering of pH if sufficient alkalinity is not present.

Oxidation does not take place alone. Synthesis also occurs, resulting in the growth of additional organisms. Synthesis may be expressed by [3] :

$$15\,\text{CO}_2 + 13\,\text{NH}_4^+ \xrightarrow{\hspace{2cm}} 10\,\text{NO}_2^- + 3\,\text{C}_5\text{H}_7\text{NO}_2 + 23\,\text{H}^+ + 4\,\text{H}_2\text{O} \quad (4.8)$$
$$\textit{Nitrosomonas}$$

and

$$5\,\text{CO}_2 + \text{NH}_4^+ + 10\,\text{NO}_2^- + 2\,\text{H}_2\text{O} \xrightarrow{\hspace{2cm}} 10\,\text{NO}_3^- + \text{C}_5\text{H}_7\text{NO}_2 + \text{H}^+ (4.9)$$
$$\textit{Nitrobacter}$$

A knowledge of cell yield will permit the combination of the oxidation and synthesis reactions. Literature values indicate cell yields of from 0.04 to 0.29 mg VSS/mg NH_4-N for *Nitrosomonas* and from 0.02 to 0.084 mg VSS/mg NH_4-N for *Nitrobacter* [7-10]. Using an assumed cell yield of 0.15 mg VSS/mg NH_3-N for *Nitrosomonas* and 0.02 mg VSS/mg NO_2-N for *Nitrobacter*, we can show that [3] :

$$55\,\text{NH}_4^+ + 76\,\text{O}_2 + 109\,\text{HCO}_3^- \xrightarrow{\hspace{3cm}}$$

$$\textit{Nitrosomonas}$$
$$\text{C}_5\text{H}_7\text{NO}_2 + 54\,\text{NO}_2^- + 57\,\text{H}_2\text{O} + 104\,\text{H}_2\text{CO}_3 \qquad (4.10)$$

and

$$400\,\text{NO}_2^- + \text{NH}_4^+ + 4\,\text{H}_2\text{CO}_3 + \text{HCO}_3^- + 195\,\text{O}_2 \xrightarrow{\hspace{3cm}}$$

$$\textit{Nitrobacter}$$
$$\text{C}_5\text{H}_7\text{NO}_2 + 3\,\text{H}_2\text{O} + 400\,\text{NO}_3^- \qquad (4.11)$$

Combining and simplifying Equations 4.10 and 4.11 yields:

$$NH_4 + 1.83\ O_2 + 1.98\ HCO_3^- \longrightarrow 0.021\ C_5H_7NO_2 + 1.04\ H_2O$$

$$+ 0.98\ NO_3^- + 1.88\ H_2CO_3 \qquad (4.12)$$

The oxygen requirement for nitrification may be calculated from Equation 4.12 to be approximately 4.3 mg O_2/mg NH_4-N. This value agrees well with oxygen requirements found in the literature [3,8,10].

From Equation 4.12 it is also possible to calculate the alkalinity consumed during nitrification. The alkalinity consumed is 7.14 mg of alkalinity consumed as $CaCO_3$ per mg NH_4-N oxidized. This value matches very well with values found in the literature [3,8,10,11]. Further discussion of alkalinity is contained in the section on kinetics.

KINETICS OF NITRIFICATION

In considering the kinetics of nitrification, we will use the Monod model for limiting substrate [12-14]. As previously discussed in Chapter 2, several other models have been developed [15-19] which yield similar results at steady-state conditions. When highly variable influent ammonia concentrations exist, the use of these other models will permit prediction of effluent ammonia concentrations as a function of influent nitrogen concentration.

Using the Monod model with NH_4-N as the rate-limiting nutrient, we get:

$$\mu_n = (\mu_n)_{max}\ \frac{N}{K_n + N} \qquad (4.13)$$

where μ_n = growth rate of nitrifiers, day^{-1}
$(\mu_n)_{max}$ = maximum growth rate of nitrifiers, day^{-1}
N   = NH_4-N concentration, mg/L
K_n = half-velocity constant for nitrification, mg/L

It has been shown that the energy yield from the conversion of ammonia to nitrite is approximately three times the energy yield from the conversion of nitrite to nitrate [20]. Thus, three times as much nitrate as nitrite must be processed to produce the same energy yield. This may be used to explain why nitrite is rarely found in significant concentrations in biological treatment systems [3,21]. For this reason, it is clear that the rate-limiting step in nitrification is the conversion of ammonia to nitrite by *Nitrosomonas*. Thus, in Equation 4.13, the growth rates are taken as the growth rates of *Nitrosomonas*, and the values of N and K_n represent concentrations of NH_4-N.

It has been shown [3,7,8,21-24] that the half-velocity constant (K_n) is less than 1 mg/L at 20°C for both *Nitrosomonas* and *Nitrobacter*. This, combined with energy yields, indicates that the growth rate of *Nitrobacter* is much higher than the growth rate of *Nitrosomonas*. Therefore, it may be stated that (1) nitrite will not accumulate in significant amounts; and (2) the overall growth of nitrifiers may be modeled using the conversion of ammonia to nitrite as rate-limiting.

The rate of oxidation of ammonia can be related to the growth rate of *Nitrosomonas* by Equation 4.14:

$$q_n = \frac{\mu_n}{Y_n} = \frac{(\mu_n)_{max}}{Y_n} \left(\frac{N}{N + K_n} \right) \tag{4.14}$$

where

q_n = ammonia oxidation rate, mg NH_4-N oxidized/mg VSS/day

$(q_n)_{max}$ = maximum ammonia oxidation rate, mg NH_4-N oxidized/ mg VSS/day

Y_n = growth yield coefficient, mg *Nitrosomonas* grown/mg NH_4-N oxidized

It should be noted that K_n is very low under most conditions. The rate of nitrification may be taken as zero-order with respect to substrate, although it is first-order with respect to microbial concentration.

Environmental Effects on Kinetics

Nitrification kinetics are strongly affected by the environmental conditions within the bioreactor. The environmental conditions of concern include: temperature, pH, alkalinity, dissolved oxygen and toxic or inhibitory compounds.

Temperature

Temperature affects both the value of the half-velocity constant K_n and the growth rate of *Nitrosomonas* (μ_n). Downing and co-workers [25,26] have estimated the effect of temperature on the half-velocity constant, K_n:

$$K_n = 10^{0.051T - 1.158} \tag{4.15}$$

where K_n = half-velocity constant, mg/L as N
$\quad\quad$ T = temperature,°C

A review of the literature [3,24,27,28] indicates good agreement of experimental results with Equation 4.15.

Downing and co-workers [25,26] have also estimated the maximum growth rates of *Nitrosomonas* as a function of temperature. The equation developed for pure cultures shows excellent agreement with experimental results [3]:

$$(\mu_n)_{max} = 0.47e^{0.098(T - 15)} \tag{4.16}$$

where $(\mu_n)_{max}$ = maximum growth rate of *Nitrosomonas*, day^{-1}
$\quad\quad\quad$ T = temperature,°C

Recent work by Charley et al. [27] and Quinlan [28] has shown that the optimum temperature for nitrification is approximately 15°C and that growth rates for *Nitrosomonas* and *Nitrobacter* are affected differently by temperature change. At elevated temperatures *Nitrosomonas* growth rates are accelerated while *Nitrobacter* growth rates are depressed [28]. This leads to potentially inhibitory accumulation of nitrite in nitrifying systems.

pH

It is clear that the growth rate of *Nitrosomonas*, and therefore the nitrification rate, is affected by pH. A review of the literature [3] indicates a wide range of pH values reported as optimum. There is general agreement that the optimum pH for nitrification lies between 7.2 and 8.8. As the pH of the system is reduced below 7.2, the growth rate of *Nitrosomonas* is reduced. It should be noted that the effect of pH change in acclimated cultures is considerably less than the effect in nonacclimated cultures.

Downing and co-workers [25,26] have proposed that for pH values less than 7.2:

$$\mu_n = (\mu_n)_{max} [1 - 0.833(7.2 - pH)] \tag{4.17}$$

At a pH of 6.8, Equation 4.17 yields a growth rate for *Nitrosomonas* equal to 67% of the maximum growth rate.

Oxidation ditches, being essentially completely mixed reactors, are resistant to sudden pH changes due to influent wastewater pH changes. The large volume of the oxidation ditch provides significant buffering and dilution capacity to resist pH changes within the reactor.

Inspection of Equation 4.12 reveals that significant levels of hydrogen ions (as H_2CO_3) are produced during nitrification. Thus, even with wastewater having a pH greater than 7.0, the pH within the oxidation ditch will drop below optimum unless sufficient alkalinity remains after nitrification to provide the required buffering capacity.

The pH in an operating oxidation ditch may be obtained from:

$$pH = pK_1 - \log \frac{[H_2CO_3]}{[HCO_3^-]} \qquad (4.18)$$

where pK_1 = - log of the ionization constant for H_2CO_3

A value for $[H_2CO_3]$ may be approximated from Henry's law while accounting for the production of carbon dioxide during biological processes.

A more practical method of ensuring proper pH conditions in an oxidation ditch for nitrification is to ensure that residual alkalinity in the oxidation ditch after nitrification is greater than 100 mg/L as $CaCO_3$.

Alkalinity

Alkalinity serves as a buffer to maintain pH near neutral in natural water systems as well as oxidation ditches. Equation 4.12 also indicates the role of alkalinity in nitrification. For practical purposes, it may be assumed that 7.14 g of alkalinity is destroyed per g of NH_3 -N oxidized [3,11].

Alkalinity is produced during the oxidation of carbonaceous BOD at long solids retention times. This may be seen from the equations developed by McCarty [4]. The alkalinity produced or destroyed during carbonaceous BOD removal is presented as a function of solids retention time in Figure 4.1 for various values of growth yield and endogenous decay coefficients. As may be seen from Figure 4.1, as much as 0.3 mg of alkalinity may be produced per mg BOD_5 removed at the long solids retention times typical of oxidation ditches. For an influent BOD_5 of 220 mg/L this represents approximately 60 mg/L of alkalinity produced, which is sufficient alkalinity for the oxidation of approximately 9 mg/L of NH_3-N.

Dissolved Oxygen

There is clear evidence [3] that dissolved oxygen concentration affects the rate of nitrification. Most investigators agree that the use of a Monod-type relationship

$$\mu_n = (\mu_n)_{max} \left[\frac{O_2}{K_{O_2} + O_2} \right] \qquad (4.19)$$

where O_2 = oxidation ditch dissolved oxygen concentration, mg/L
            K_{O_2} = half-velocity constant for oxygen, mg/L

will adequately describe the effect of dissolved oxygen on nitrification rate. Many values of K_{O_2} have been proposed, ranging from 0.25 to 2.46 mg/L. The investigations leading to these values have been summarized by Stenstrom and Poduska [29].

The U.S. EPA [3] and others [10] recommend the use of 1.3 mg/L for K_{O_2}. At this value for K_{O_2}, dissolved oxygen concentration stops significantly affecting nitrification rates when the dissolved oxygen concentration exceeds 2 mg/L. It has been shown [3,29] that complete nitrification can be achieved at dissolved oxygen concentrations less than 2.0 mg/L but excessively long solids retention times may be required. For this reason, it is common to design oxidation ditches to maintain dissolved oxygen at or above 2.0 mg/L in the aerobic portion of the ditch.

Inhibitory Compounds

As with other biological systems, nitrification processes are subject to inhibition or toxicity from a variety of compounds. Of particular concern are heavy metals, phenols, thiourea and its derivatives, and free ammonia [30,31]. One investigator [27] has found dissolved oxygen to be inhibitory at high concentrations. The completely mixed nature of oxidation ditches serves to minimize toxic and inhibitory effects. However, if it is suspected that toxic or inhibitory compounds may be present in the wastewater, bench- or pilot-scale studies to screen for effects on nitrification rate are strongly recommended.

Combined Effect of Environmental Conditions on Nitrification Kinetics

We have shown that environmental conditions may have a significant impact on nitrification kinetics. Each environmental parameter has been considered independently. These effects may be combined as the product of several Monod-type equations [32]. By combining Equations 4.13, 4.17 and 4.19 we get:

$$\mu_n = (\mu_n)_{max} \left(\frac{N}{K_n + N} \right) \left(\frac{O_2}{K_{O_2} + O_2} \right) [1 - 0.833 (7.2 - pH)] \qquad (4.20)$$

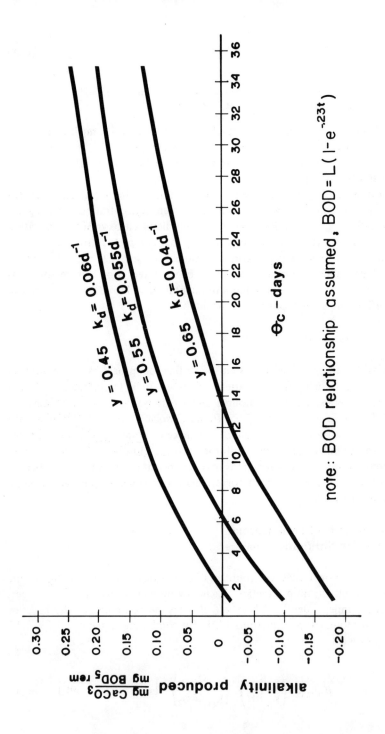

Figure 4.1. Alkalinity produced (required) during carbonaceous BOD removal.

We can further expand Equation 4.20 by substituting Equation 4.15 for K_n and Equation 4.16 for $(\mu_n)_{max}$. Performing these substitutions yields:

$$\mu_n = \left[0.47e^{0.098(T-15)}\right] \times \left[\frac{N}{N + 10^{0.051T-1.158}}\right] \times \left[\frac{O_2}{K_{O_2} + O_2}\right]$$

$$\times \quad [1 - 0.833(7.2 - pH)] \qquad (4.21)$$

Thus, with Equation 4.21 we can calculate, for any temperature, pH, reactor ammonia content and dissolved oxygen, the growth rate of *Nitrosomonas*.

The rate of nitrification (q_n) may then be expressed in terms of μ_n by Equation 4.14:

$$q_n = \frac{\mu_n}{Y_n} \qquad (4.14)$$

APPLICATION OF NITRIFICATION KINETICS TO OXIDATION DITCH DESIGN

The kinetics of nitrification are used in the design of nitrification within oxidation ditches. Oxidation ditches are generally designed as extended aeration systems in which nitrification occurs as well as stabilization of the biological solids. The nitrification kinetics which have been discussed may be used to:

1. determine the design solids retention time under the most adverse conditions of pH, dissolved oxygen and temperature
2. determine the minimum allowable hydraulic retention time for carbonaceous BOD removal as well as for nitrification
3. determine sludge wasting rate

Our discussions to this point have treated nitrification kinetics as if they were occurring in a pure culture of nitrifiers. This is clearly not the case in an oxidation ditch. The wastewater entering an oxidation ditch contains large quantities of biodegradable organic materials. Thus, the microbial population within the ditch consists predominantly of the faster growing heterotrophs with a small fraction of nitrifiers.

The cell yield of the heterotrophic organisms is much greater than that for nitrifiers. Thus, if the growth rate of heterotrophs (μ_h) becomes too much greater than the growth rate of the nitrifiers (μ_n), the nitrifiers will be washed out of the system. Therefore, to maintain a nitrifying system, it is necessary that:

$$\mu_n \geqslant \mu_h \tag{4.22}$$

Since we know that:

$$\theta_c = \frac{1}{\mu} \tag{4.23}$$

we can say that:

$$\theta_c^d \geqslant \theta_c^m = \frac{1}{\mu_n} \tag{4.24}$$

where θ_c^d = design solids retention time, days

θ_c^m = minimum solids retention time for nitrification at design environmental conditions, days

In the oxidation ditch, the environmental conditions of temperature, dissolved oxygen, ammonia levels and pH will determine μ_n and therefore the minimum solids retention time (θ_c^m). Since θ_c^m is fixed by environmental conditions, in order to control the ditch so that nitrification can occur, we must adjust μ_h or θ_c^d.

We have previously shown that:

$$\mu_h = \frac{1}{\theta_c^d} = YU - k_d \tag{2.35}$$

and that:

$$U = \frac{S_o - S}{\theta X} \tag{2.36}$$

Looking at the above equations, it is clear that since Y and k_d are system constants which we cannot control, the only way to alter μ_h or θ_c^d is to modify U. This can be accomplished by altering reactor volume (and therefore hydraulic retention time, θ) or the concentration of MLVSS, X.

In an oxidation ditch, there is no method to selectively waste either heterotrophs or nitrifiers from the system. Therefore, the cell wasting rate and thus the solids retention times and growth rates for nitrifiers and heterotrophs must be the same.

Safety Factor

Lawrence and McCarty [33] introduced the concept of a safety factor to the design of nitrification systems. The safety factor is defined as:

$$S.F. = \frac{\theta_c^d}{\theta_c^m} \qquad (4.25)$$

They recommended a conservative safety factor to account for unexpected changes in environmental conditions and/or the presence of toxic or inhibitory compounds.

The concept of safety factor can also be used to provide sufficient system capacity to handle peaks in influent nitrogen load. A nonsteady-state mass balance can be used [3] to determine the effect of peak influent loads on effluent NH_3-N levels at various safety factors. The effect of safety factor on effluent NH_3-N as a result of influent peak loads may be seen in Figure 4.2.

The application of nitrification kinetics to the design of oxidation ditches is best shown by example.

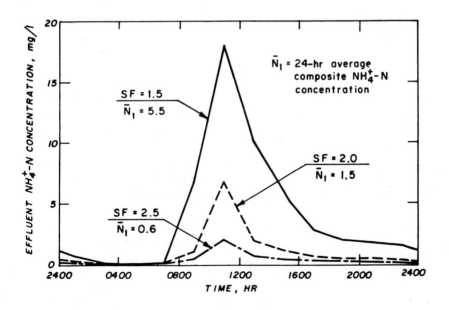

Figure 4.2. Effect of safety factor on effluent ammonia [3].

Example 4.1. Design an oxidation ditch to treat 1 MGD of wastewater having the following characteristics:

- BOD_5 = 200 mg/L
- TSS = 200 mg/L
- TKN = 30 mg/L
- alkalinity = 350 mg/L as $CaCO_3$
- minimum temperature = $10^{\circ}C$

The effluent is to be completely nitrified and is to contain not more than 5.0 mg/L of soluble BOD_5.

In formulating the design we will make certain assumptions:

1. We will ignore that portion of the TKN used for synthesis and assume that all TKN removed is oxidized.
2. A safety factor of 3.0 will be assumed to handle peak flows.
3. The dissolved oxygen in the oxidation ditch will be assumed to be 2.0 mg/L.

First we will check to see if sufficient alkalinity is present to maintain the pH.

Assuming that all of the influent TKN is oxidized, the alkalinity required will be:

$$\text{alkalinity required} = 30 \text{ mg/L TKN} \times 7.14 \ \frac{\text{mg alk.}}{\text{mg TKN}} = 214 \text{ mg/L as } CaCO_3$$

The residual alkalinity is then:

$$350 \text{ mg/L} - 214 \text{ mg/L} = 136 \text{ mg/L as } CaCO_3$$

With the high-turbulence aerators used in oxidation ditches, good stripping of the carbon dioxide produced may be expected, and a residual alkalinity of 136 mg/L as $CaCO_3$ should be sufficient to maintain a pH greater than 7.2.

Now, calculate the maximum growth rate of *Nitrosomonas* at $10^{\circ}C$, a dissolved oxygen of 2.0 mg/L and a pH greater than 7.2. It is still assumed that nitrogen is not rate-limiting. Thus, under these conditions:

$$(\mu_n)_{max} = 0.47 e^{0.098(T-15)} \left(\frac{O_2}{O_2 + K_{O_2}} \right)$$

Substituting:

$$T = 10^{\circ}C$$

$$O_2 = 2 \text{ mg/L}$$

and assuming:

$$K_{O_2} = 1.3 \text{ mg/L}$$

we get:

$$(\mu_n)_{max} = 0.47e^{0.098(10-15)} \left(\frac{2.0}{2.0 + 1.3} \right)$$

$$(\mu_n)_{max} = 0.175 \text{ day}^{-1}$$

and:

$$\theta_c^m = \frac{1}{(\mu_n)_{max}} = \frac{1}{0.175} = 5.71 \text{ days}$$

then:

$$\theta_c^d = \text{S.F.} \times \theta_c^m = 3.0 \times 5.71 = 17.13 \text{ days}$$

say:

$$\theta_c^d = 17 \text{ days}$$

This now sets the design growth rate for both nitrifiers and heterotrophs:

$$\mu_n = \mu_h = \frac{1}{\theta_c^d} = \frac{1}{17} = 0.0588 \text{ day}^{-1}$$

We can now determine the effluent NH_3-N from:

$$\mu_n = (\mu_n)_{max} \left[\frac{N}{K_n + N} \right]$$

and:

$$K_n = 10^{0.051T - 1.158}$$

$$K_n = 10^{0.051 \times 10 - 1.158} = 0.22 \text{ mg/L}$$

Substituting:

$$0.0588 = 0.175 \left[\frac{N}{0.22 + N} \right]$$

Solving:

$$N = 0.11 \text{ mg/l}$$

which is essentially complete nitrification.

We can now calculate the organic removal rate. We know that:

$$\mu_n = \mu_h = 0.0588 \text{ day}^{-1} = YU - k_d$$

Assuming:

$$Y = 0.55 \text{ mg VSS/mg BOD}_5$$

and

$$k_d = 0.05 \text{ day}^{-1}$$

We can calculate U:

$$U = \frac{\mu_h + k_d}{Y} = \frac{0.0588 + 0.05}{0.55} = 0.198 \frac{\text{lb BOD}_5 \text{ rem.}}{\text{lb MLVSS-day}}$$

We then can determine the basin volume and hydraulic retention times required. We will remove:

$$1 \text{ MGD} \times (200 - 5) \text{ mg/L BOD}_5 \times 8.34 = 1626 \text{ lb BOD}_5/\text{day}$$

Thus:

$$\text{lb MLVSS req.} = \frac{1626 \text{ lb BOD}_5/\text{day}}{0.198 \dfrac{\text{lb BOD}_5 \text{ rem.}}{\text{lb MLVSS-day}}} = 8212 \text{ lb MLVSS}$$

If we now assume that we will maintain 2500 mg/L MLVSS, the required oxidation ditch volume is:

$$V = \frac{8212 \text{ lb MLVSS}}{2500 \text{ mg/L} \times 8.34} = 0.394 \text{ MG}$$

and the hydraulic retention time is:

$$\theta = \frac{V}{Q} = \frac{0.394 \text{ MG}}{1 \text{ MGD}} \times \frac{24 \text{ hr}}{\text{day}} = 9.5 \text{ hr}$$

DENITRIFICATION

The heterotrophic organisms responsible for carbonaceous removal are capable of using nitrate and nitrite as an oxygen source under anoxic conditions. This process, often termed nitrate respiration, results in the reduction of nitrate and nitrite to nitrogen gas. In oxidation ditches, provision of an anoxic zone and the use of the organics in the wastewater as the carbon source result in a significant degree of denitrification in the ditch without the addition of chemicals.

Biochemistry of Denitrification

The early work in denitrification of wastewaters was done using methanol as the carbon source. Using methanol, McCarty et al. [34] proposed the following equations to describe denitrification:

Overall nitrate removal:

$$NO_3^- + 1.08\ CH_3OH + 0.24\ H_2CO_3 = 0.056\ C_5H_7NO_2 + 0.47\ N_2 + 1.67\ H_2O$$

$$+ HCO_3^- \qquad (4.26)$$

Overall nitrite removal:

$$NO_2^- + 0.53\ H_2CO_3 + 0.67\ CH_3OH = 0.04\ C_5H_7NO_2 + 0.48\ N_2 + 1.23\ H_2O$$

$$+ HCO_3^- \qquad (4.27)$$

Overall deoxygenation:

$$O_2 + 0.93\ CH_3OH + 0.56\ NO_3^- = 0.056\ C_5H_7NO_2 + 1.04\ H_2O + 0.59\ H_2CO_3$$

$$+ 0.056\ HCO_3^- \qquad (4.28)$$

From Equations 4.26, 4.27 and 4.28 we can show that:

1. A total of 3.47 mg of alkalinity is produced per mg (NO_3-N + NO_2-N) removed.
2. The growth yield of organisms may be described by:

$$Y_{dn} = 0.53\ NO_3\text{-}N + 0.32\ NO_2\text{-}N + 0.19\ DO \qquad (4.29)$$

where Y_{dn} = growth yield coefficient mg cells/mg (NO_3-NO_2-N)
NO_3-N = nitrate concentration, mg/L as N

$$NO_2\text{-}N \quad = \text{nitrate concentration, mg/L as N}$$
$$DO \quad = \text{dissolved oxygen concentration, mg/L}$$

3. The methanol required may be calculated from:

$$CH_3OH = 2.47\ NO_3^-\text{-}N + 1.53\ NO_2^-\text{-}N + 0.87\ DO \tag{4.30}$$

It is more useful, however, to express the organic substrate in terms of COD rather than methanol, since methanol is rarely used, but the use of wastewater as an organic source is common. Thus, Equation 4.30 can be rewritten as:

$$COD_{req.} = 3.71\ NO_3^-\text{-}N + 2.3\ NO_2^-\text{-}N + 1.3\ DO \tag{4.31}$$

where $COD_{req.}$ = required biodegradable COD, mg/L

Environmental Conditions

The environmental conditions of concern for denitrification are pH, temperature, organic substrate, and inhibitory or toxic compounds. Of these we need not be concerned with the latter in denitrification. Compounds which are toxic or inhibitory to the denitrification process will be significantly more of a problem in carbonaceous BOD removal or nitrification and must be dealt with at those stages.

Temperature

Temperature affects the rate of denitrification as it does all biochemical processes. Sutton et al. [35] found that the rate of denitrification fit the conventional Arrhenius relationship. The Arrhenius relationship is usually simplified for the range of temperatures encountered in wastewater treatment to:

$$K_T = K_{20}\theta^{T-20} \tag{4.32}$$

where K_T = rate constant at temperature T, time^{-1}
K_{20} = rate constant at 20°C, time^{-1}
θ = constant
T = temperature, °C

Values of θ have been reported between 1.06 and 1.12 [35-37].

pH and Alkalinity

Several studies on the effect of pH on the rate of denitrification [3] all show agreement that the rate of denitrification is optimized in the pH range of 7.0 to 7.5. It is also clear from these studies that outside of the pH range of 6.0 to 8.0 the rate of denitrification is significantly reduced.

As shown in Equations 4.26, 4.27 and 4.28, alkalinity is produced during denitrification. This helps maintain the pH in the desired range and replaces some portion of the alkalinity destroyed during nitrification. The predicted 3.57 g of alkalinity produced per g of N reduced has not been seen in actual fieldwork. Two studies [38,39] have found values of 2.95 and 2.89 g/g, respectively. The U.S. EPA [3] recommends the use of 3 g/g as a good engineering design value.

Organic Substrate

Unlike nitrification, denitrification depends on heterotrophic organisms. Denitrification results when, during the metabolism of organic material, oxidized forms of nitrogen (NO_3^-, NO_2^-) are used as the source of oxygen. In order for denitrification to proceed at its maximum rate, biodegradable organic material must be present in excess.

It has been shown [37] that when methanol is used as the carbon source a Monod type expression may be used to express the effect of organic substrate availability.

$$\mu_{dn} = (\mu_{dn})_{max} \frac{C}{K_c + C} \tag{4.33}$$

where
μ_{dn} = denitrifier growth rate, day^{-1}
$(\mu_{dn})_{max}$ = maximum denitrifier growth rate, day^{-1}
C = organic substrate concentration, mg/L
K_c = half-velocity constant for organic substrate, mg/L

When methanol was used, it was found [3] that K_c was very low, on the order of 0.1 mg/L as methanol.

The Monod model has limited usefulness, however, since most systems and all oxidation ditch systems are currently designed using wastewater as the carbon source. Thus, evaluating K_c becomes extremely difficult and is specific for each waste. The subject of organic substrate availability is covered in some detail by Beer [40].

It is clear that a significantly higher denitrification rate may be expected when ample biodegradable organic substrate is available than when purely endogenous respiration is occurring. This is borne out by the denitrification rates reported in the literature at high and low organic loadings [40-42].

Denitrification Kinetics

The usual Monod kinetics may be used to describe denitrification processes if the process is assumed to be a one-step conversion of nitrate to nitrogen gas [3]. This is a reasonable approach since it is rare that significant quantities of nitrite are present in wastewater systems.

The effect of nitrate concentration on the kinetics of denitrification may be described by:

$$\mu_{dn} = (\mu_{dn})_{max} \frac{N}{K_{dn} + N} \tag{4.34}$$

where N = nitrate concentration, NO_3-N mg/L
        K_{dn} = half-velocity constant for nitrate, NO_3-N mg/L

The value of K_n has been reported by numerous investigators to be from 0.06 to 0.16 mg/L of NO_3^--N [43,44]. Thus, at concentrations of NO_3^--N of 1 to 2 mg/L, the NO_3^--N concentration no longer affects the rate of denitrification, and denitrification may be considered to be zero-order with respect to nitrate concentration [37,40,43-45].

As was done earlier, the rate of denitrification can be related to the growth rate of denitrifiers by:

$$q_{dn} = \frac{\mu_{dn}}{Y_{dn}} \tag{4.35}$$

where q_{dn} = denitrification rate, g NO_3-N removed/g MLVSS-day

       Y_{dn} = denitrifier yield constant, g VSS grown/g NO_3^--N removed

Also as done earlier, we can relate solids retention time and denitrification rate by:

$$\frac{1}{\theta_c} = Y_{dn} q_{dn} - k_d \qquad (4.36)$$

where θ_c = solids retention time, days
   k_d = decay coefficient, days^{-1}

Due to the variety of factors which affect denitrification rates, values for the constants involved are difficult to obtain. Pilot studies are always recommended. Table 4.2 presents kinetic constants available from the literature.

Oxygen Considerations

When nitrate serves at the terminal electron acceptor (oxygen source) in heterotrophic metabolism, it replaces dissolved oxygen. Thus, in calculation of the oxygen required to satisfy the BOD, credit may be taken for oxygen supplied through denitrification.

It has been estimated [3] that approximately 2.6 g of oxygen is supplied by each g of nitrate reduced.

In addition to providing oxygen, the nitrate contained in the oxidation ditch also serves to provide a buffer against peak oxygen demands. This buffer may be substantial. A 10-mg/L NO_3^--N concentration in a 1-MGD oxidation ditch provides sufficient buffer to satisfy 2168 lb of BOD.

Table 4.2. Kinetic Constants for Denitrification

q_{dn}, g NO_3-N rem./ g MLVSS-day	Y_{dn}, g VSS/ g NO_3-N rem.	K_d, /day $^{-1}$	T, °C	Organic Substrate	Reference
	0.57	0.05	10	methanol	45
	0.63	0.04	20	methanol	45
	0.67	0.02	30	methanol	45
0.16–0.90			20	methanol	46
0.24–3.8			5–27	sodium citrate	36
0.36–0.144	0.7–1.4		5–20	methanol	35
0.05–0.18		0.064	16–26	wastewater	41
0.069–0.26			15–27	wastewater	40
0.019–0.033				wastewater	42

APPLICATION OF DENITRIFICATION KINETICS
TO OXIDATION DITCH DESIGN

Oxidation ditches are uniquely suited to denitrification. After designing the ditch for carbonaceous BOD removal and nitrification, denitrification may be achieved simply by extending the length of the oxidation ditch to provide for an anoxic zone. Typically, wastewater will be introduced just upstream of the anoxic zone to provide organic substrate for denitrification. Straightening out the ditch shows schematically the zones present (Figure 4.3).

The best way to demonstrate the use of kinetics in oxidation ditch design is by example.

Example 4.2. Continuing with the Example 4.1 presented earlier in this chapter, design the anoxic section of the oxidation ditch to reduce the NO_3^--N concentration to 10 mg/L.

We will again make certain assumptions:

1. All of the influent TKN is converted to NO_3-N.
2. At effluent NO_3-N of 10 mg/L, denitrification rates are zero-order with respect to nitrate concentration.
3. We will again ignore that fraction of the nitrogen used in synthesis.

First, we will select a design denitrification rate. Referring to Table 4.1 we see that for wastewater the low end of the range of denitrification rates is approximately 20°C. We will use the low-end values since in an oxidation ditch, which approaches a completely mixed reactor, the organic content will be low—approximately that of the effluent. Therefore we will use 0.025 g/g/day at 20°C.

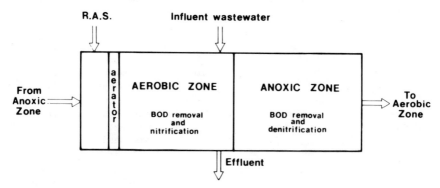

Figure 4.3. Oxidation ditch schematic.

Now, we will correct q_{dn} to $10°C$.

$$(q_{dn})_T = (q_{dn})_{20} \theta^{T-20}$$

Assuming $\theta = 1.08$:

$$(q_{dn})_{10°} = 0.025 \times 1.08^{10-20} = 0.012 \; \frac{g \; NO_3\text{-}N}{g \; MLVSS\text{-}day}$$

Thus, we can calculate hydraulic retention time required.

$$q_{dn} = 0.012 \; \frac{lb \; NO_3\text{-}N \; rem.}{lb \; MLVSS\text{-}day}$$

NO_3-N to be removed = $(30 - 10)$mg/L x 1 MGD x 8.34 = 167 lb/day

$$0.012 = \frac{167 \; lb \; NO_3\text{-}N}{day} \; x \; \frac{1}{lb \; MLVSS \; req.}$$

lb MLVSS req. = 13,917 lb

At 2500 mg/L MLVSS, the volume required:

$$V = \frac{13,917 \; lb}{2500 \; mg/L \; x \; 8.34} = 0.667 \; MG$$

Thus, hydraulic retention time required in the anoxic zone:

$$\theta = \frac{0.667 \; MG}{1 \; MGD} \; x \; 24 \; \frac{hr}{day} = 16 \; hr$$

Thus, the overall reactor (Examples 4.1 and 4.2) will be 25.5 hr hydraulic retention time, with a MLVSS of 2500 mg/L and a design of θ_c of 17 days.

REFERENCES

1. Water Pollution Control Federation, *MOP4, Chlorination of Wastewater* (Washington, DC: WPCF, 1976).
2. "Quality Criteria for Water," U.S. EPA, Washington, DC (1976).
3. "Process Design Manual for Nitrogen Control," U.S. EPA, Washington, DC (1975).
4. Comly, H. H., "Cyanosis in Infants in Well Water," *J. Am. Med. Assoc.* 129:112 (1945).
5. Walton, G., "Survey of Literature Relating to Infant Methemoglobinemia Due to Nitrate Contaminated Water," *Am. J. Public Health* 41:986 (1951).
6. "National Interim Primary Drinking Water Regulations," U.S. EPA, EPA-570/9-76-003 (1976).
7. Painter, H. A., "A Review of Literature on Inorganic Nitrogen Metabolism in Microorganisms," *Water Res.* 4(6):393 (1970).
8. Gujer, W., and D. Jenkins, "The Contact Stabilization Process—Oxygen and Nitrogen Mass Balances," SERL Report, 74, Univ. of California (1974), p. 2.
9. Stensel, H. D., and J. H. Scott, "Cost Effective Advance Biological Wastewater Treatment Systems," presented at Nevada Water Pollution Assoc. Annual Meeting (Dec., 1977).
10. Metcalf and Eddy, Inc., *Wastewater Engineering*, 2nd Edition (New York: McGraw-Hill Book Company, 1979).
11. Sherrard, J. H., "Destruction of Alkalinity in Aerobic Biological Wastewater Treatment," *J. Water Poll. Control Fed.*, 48(7):1834-1839 (1976).
12. Monod, J., *Recherches sur la Croissance des Cultures Bacteriennes* (Paris: Herman et Cie., 1942).
13. Monod, J., "La Technique of Culture Continue; Theorie et Applications," *Annals Institue Pasteur*, 79:340 (1950).
14. Monod, J., "The Growth of Bacterial Cultures," *Ann. Rev. Microbiol.* III (1949).
15. Grady, C. P. N., Jr. and D. R. Williams, "Effects of Influent Substrate Concentration on the Kinetics of Natural Microbial Populations in Continuous Cultures," *Water Res.*, 9:171 (1975).
16. Grau, P., M. Dohanyos and J. Chudoba, "Kinetics of Multicomponent Substrate Removal by Activated Sludge," *Water Res.*, 9:637 (1975).
17. Benefield, L. D., and C. W. Randall, "Evaluation of a Comprehensive Kinetic Model for the Activated Sludge Process," *J. Water Poll. Control Fed.*, 49:1636 (1977).
18. Adams, C. D., W. W. Eckenfelder and J. Hovius, "A Kinetic Model for Design of Completely Mixed Activated Sludge Treating Variable Strength Industrial Wastewaters," *Water Res.*, 9(1):37 (1975).
19. Elmaleh, S., and R. Ben Aim, "Influence sur la Cinétique Biochemique de la Concentration en Carbone Organique à l'entrée d'un Réacteur Developpement une Polyculture Microbienne en Mélange Parfait," *Water Res.*, 10:1005 (1976).
20. McCarty, P. L., "Stoichiometry of Biological Reactions," *Progress in Water Technology* (London: Pergamon Press, 1975).

21. Adams, C. E., Jr., and W. W. Eckenfelder, Jr., "Nitrification Design Approach for High Strength Ammonia Wastewaters," *J. Water Poll. Control Fed.*, 49(3):413-421 (1977).

22. Haus, R. T., and P. L. McCarty, "Nitrification with the Submerged Filter," Report prepared by the Department of Civil Engineering, Stanford University for the U.S. EPA, Research Grant No. 17010 EPM (1971).

23. Stratton, F. E., and P. L. McCarty, "Prediction of Nitrification Effects on the Dissolved Oxygen Balance of Streams," *Environ. Sci. Technol.*, 1(5):405-410 (1967).

24. Hall, E. R., and K. L. Murphy, "Estimation of Nitrifying Biomass and Kinetics in Wastewater," *Water Res.*, 14:297-304 (1980).

25. Knowles, G., A. L. Downing and M. J. Barnett, "Determination of Kinetic Constants for Nitrifying Bacteria in Mixed Culture with the Aid of an Electronic Computer," *J. Gen. Microbiol.*, 38:263 (1965).

26. Downing, A. L., and A. P. Hopwood, "Some Observations on the Kinetics of Nitrifying Activated Sludge Plants," *Schweiz. Zeitsch Hydrol.*, 26(2):271 (1964).

27. Charley, R. C., D. G. Hooper and A. G. McLee, "Nitrification Kinetics in Activated Sludge at Various Temperatures and Dissolved Oxygen Concentrations," *Water Res.*, 14:1387-1396 (1980).

28. Quinlan, A. V., "The Thermal Sensitivity of Nitrification as a Function of the Concentration of Nitrogen Substrate," *Water Res.*, 14: 1501-1507 (1980).

29. Stenstrom, M. K., and R. A. Poduska, "The Effect of Dissolved Oxygen Concentration on Nitrification," *Water Res.*, 14:643-649 (1980).

30. Wood, L. B., B. J. E. Hurley and P. J. Matthews, "Some Observations on the Biochemistry and Inhibition of Nitrification," *Water Res.*, 15:543-551 (1981).

31. Neufeld, R. D., A. J. Hill and D. O. Adexoya, "Phenol and Free Ammonia Inhibition to *Nitrosomonas* Activity," *Water Res.*, 14:1695-1703 (1980).

32. Chen, C. W., "Concepts and Utilities of Ecological Model," *J. San. Eng. Div., ASCE*, 96(SA5):1085-1097 (1970).

33. Lawrence, A. W., and P. L. McCarty, "A Unified Basis for Biological Treatment Design and Operation," *J. San. Eng. Div., ASCE*, 96(SA3): 757 (1970).

34. McCarty, P. L., L. Beck and P. St. Amant, "Biological Denitrification of Wastewaters by Addition of Organic Materials," *Proceedings of the 24th Industrial Waste Conference*(W. Lafayette, IN: Purdue University, 1969).

35. Sutton, P. M., K. L. Murphy and R. N. Dawson, "Low-Temperature Biological Denitrification of Wastewater," *J. Water Poll. Control Fed.*, 47(1):122-134 (1975).

36. Dawson, R. N., and K. L. Murphy, "The Temperature Dependency of Biological Denitrification," *Water Res.*, 6:71 (1972).

37. Stensel, H. D., "Biological Kinetics of the Suspended Growth Denitrification Process," PhD Thesis, Cornell Univ., Ithaca, NY (1970).

38. Jeris, J. S., and R. W. Owens, "Pilot Scale High Rate Denitrification at Nassau County, NY," presented at the Winter Meeting of the New York Water Pollution Control Association (1974).

39. Horstkotte, G. A., D. G. Niles, D. S. Parker and D. H. Caldwell, "Full Scale Testing of a Water Reclamation System," *J. Water Poll. Control Fed.*, 46(1):181-197 (1974).

40. Beer, C., "A Study of Nitrate Respiration in the Activated Sludge Process," U.S. EPA, EPA-600/2-80-154 (1980).

41. Heidman, J. A., "Sequential Nitrification-Denitrification in a Plug Flow Activated Sludge System," U.S. EPA, EPA 600/2-79-157 (1979).

42. Bishop, D. F., J. A. Heidman, and J. B. Steinberg, "Single-Stage Nitrification-Denitrification," *J. Water Poll. Control Fed.*, 48(3):520-532 (1976).

43. Moore, S. F., and E. D. Schroeder, "The Effect of Nitrate Feed Rate on Denitrification," *Water Res.*, 5:445-452 (1971).

44. Requa, D. A., and E. D. Schroeder, "Kinetics of Packed Bed Denitrification," *J. Water Poll. Control Fed.*, 45(8):1696-1707 (1973).

45. Stensel, H. D., R. C. Loehr and A. W. Lawrence, "Biological Kinetics of the Suspended Growth Denitrification," *J. Water Poll. Control Fed.*, 45(2):249-261 (1973).

46. Moore, S. F., and E. D. Schroeder, "An Investigation of the Effects of Residence Time on Anaerobic Bacterial Denitrification," *Water Res.*, 4:685-694 (1970).

CHAPTER 5

OXYGEN REQUIREMENTS AND TRANSFER

MASS TRANSFER THEORY

In a system containing a gas and a liquid in intimate contact, an exchange of mass takes place between the liquid and the gas. Two theories have been advanced to explain the mechanism of the exchange. These are the penetration theory and the film theory.

Mass Transfer

The penetration theory hypothesizes that eddies originating in the turbulent bulk liquid migrate to the gas-liquid interface, where they are exposed briefly to the gas before being displaced by other eddies arriving at the interface. Conversely, small eddies of gas may exist temporarily in the liquid prior to dissolution. During the brief residence of eddies in the liquid at the interface, eddies absorb molecules from the gas. On return of the eddies to the liquid bulk, molecules are distributed by turbulence. The rate of mass transfer is considered to be a function of diffusivity, concentration gradient and surface renewal. The penetration theory holds considerable promise in describing mass transfer operations, especially in regard to aeration devices producing varying degrees of interfacial turbulence.

The film theory, although more questionable from a theoretical standpoint, is at present of greater practical value than the penetration theory. The film theory is based on a physical model in which two fictitious films exist at the gas-liquid interface—one liquid and one gas. These films are considered to be stagnant and furnish all resistance to mass transfer. They are thought of as persistent regardless of the level of turbulence present in the gas and the liquid; the turbulence serves only to reduce the film thickness. The imaginary transfer mechanism is depicted in Figure 5.1.

If gas is transferred to a liquid which is unsaturated with respect to the gas, the gas molecules are first transferred to the outer boundary of the gas film through the combined processes of mixing and diffusion. The molecules then diffuse across the stagnant gas film to the gas-liquid interface where they dissolve in the liquid film. The dissolved gas then diffuses through this stagnant film to the boundary between the film and the bulk liquid phase, from which it is transported throughout the bulk liquid phase by mixing. The same mechanism in reverse prevails when gases are released from a supersaturated

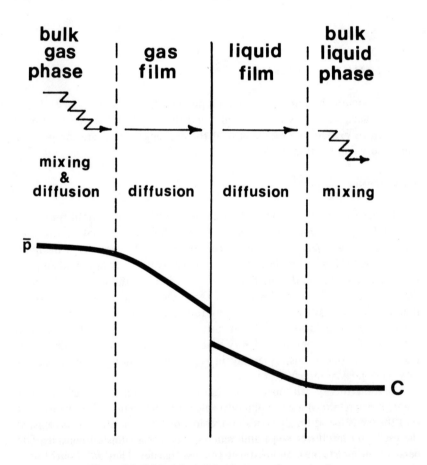

Figure 5.1. Schematic sketch of gas transfer mechanism.

solution. Since transfer by diffusion is slow compared to that by mixing, the rate of one phase to another is considered to be controlled by the stagnant films. Also, since the diffusion in the gas film proceeds at a much higher rate than in the liquid film, liquid film diffusion is assumed to be the rate-controlling step in mass transfer.

Equilibrium Relationships

If a fluid contains a component that is not homogeneously distributed, such that concentration gradients exist, forces act to transfer mass in such a manner as to minimize these concentration differences. For a gas and liquid in contact, molecules of the gas will transfer into or out of solution until the rate of transfer into solution equals the rate of transfer out of solution. The concentration of gas solute at this point is called the saturation concentration and is a function of the liquid temperature, the concentration of other solutes present, and the partial pressure of the gas phase in contact with the liquid. If the amount of gas solute is reduced, gas will transfer into the liquid at a rate which is a function of the difference between the saturation concentration and the actual solute concentration. For the purposes of gas-liquid transfer dynamics, the partial pressure of a gas and its equilibrium saturation concentration are correctly related by Henry's law.

The solubilities of gases (equilibrium concentration) in liquids vary widely. For gases of low or moderate solubility that do not react chemically with the solvent, a quantity of gas which will dissolve at a given temperature can be determined by Henry's law (Equation 5.1).

$$\bar{p} = HX \tag{5.1}$$

where $\bar{p}$ = partial pressure of the gas in equilibrium with the liquid, atm
  H = Henry's law constant, atm
  X = mol fraction of gas in equilibrium with the liquid

OXYGEN TRANSFER MODEL

Oxygen is only slightly soluble in water. As such, the rate of diffusion of oxygen molecules through the stagnant liquid film is slow with respect to the other mass transfer steps and can be taken as a rate-limiting step. The rate of transfer is proportional to the concentration gradient across the liquid film determined by the liquid film thickness, δ, and the difference between the saturation dissolved oxygen level according to Henry's law at the interface and the dissolved oxygen level in the bulk mass of the liquid.

$$\frac{dC}{dt} = K_{La} \ (C^* - C) \tag{5.2}$$

where
- C = oxygen concentration, mg/L
- t = time
- K_{La} = overall oxygen mass transfer coefficient, L/T
- C^* = equilibrium (saturation) oxygen concentration corresponding to p

K_{La} is the product of mass transfer velocity, K_L, and the interfacial oxygen transfer area per unit volume of solution. K_L in turn is directly proportional to diffusivity as expressed by the liquid film coefficient, D_L, and inversely proportional to the liquid film thickness, δ.

$$K_{La} = K_L \ \frac{A}{V} \tag{5.3}$$

where
- A = area across which mass transfer occurs, ft^2
- V = volume of liquid, ft^3

and

$$K_L = \frac{D_L}{\delta} \tag{5.4}$$

It should be noted that the terms in Equations 5.2 through 5.4 exhibit considerable spacial and temporal variation within an aeration basin. Mathematical models based on the film theory and Equation 5.2 historically have been used to predict performance and analyze test data for both surface and subsurface aeration systems. Baillod [1,2] and others [3-6] have shown that these spatial and temporal variations can be accommodated within the film theory and its resulting mathematical models.

The general model for oxygen transfer in a dispersed gas-liquid system is given by:

$$\left(\begin{array}{c} \text{rate of mass transfer} \\ \text{per unit volume of} \\ \text{liquid} \end{array} \right) = \left(\begin{array}{c} \text{volumetric} \\ \text{mass transfer} \\ \text{coefficient} \end{array} \right) \times \left(\begin{array}{c} \text{driving} \\ \text{gradient} \end{array} \right) \tag{5.5}$$

expressed in terms of the overall liquid-phase volumetric mass transfer coefficient. This relationship becomes:

$$\hat{W} = \hat{K}_L \hat{a} (\hat{C}^* - \hat{C}) \tag{5.6}$$

where W = oxygen transfer rate per unit volume of clean water, $ML^{-3}T^{-1}$
  a = interfacial surface area per unit volume, L^{-1}

The caret superscript is employed to indicate the local nature of the quantities.

In applying Equation 5.6 to an oxygen transfer system, it must be remembered that this relationship applies to a particular point. In general, the total rate of oxygen transfer is of interest and can be evaluated by integrating the local transfer rate over the volume, V.

$$WV = \int \hat{W} dV = \int \hat{K}_L \hat{a} (\hat{C}^* - \hat{C}) dV \tag{5.7}$$

Clearly, K_L, a, C^* and C may vary over time. Different assumptions made concerning the spatial-temporal variations of these quantities constitute the main difference between the various models applied to oxygen transfer systems.

In a submerged aeration system, C^* may vary with dpeth because of progressive decreases in both hydrostatic pressure and oxygen mol fraction as gas bubbles move upward. Many studies have modeled submerged aeration based on the assumptions that:

1. The volumetric transfer coefficient, K_{La}, is constant over the tank volume.
2. Good mixing exists so that C is uniform over the tank volume.
3. Oxygen is the only gas transferred.

These assumptions allow Equation 5.7 to be integrated over the tank volume to give:

$$WV = K_{La} V (C^* - C) \tag{5.8}$$

where C^* = effective average dissolved oxygen saturation concentration defined by:

$$C^* = \frac{\int C^* dV}{V} \tag{5.9}$$

Baillod [1] has shown that even though most aeration systems are comprised of an aeration zone and an accumulation zone where oxygen transfer does not take place, Equation 5.8 can be used in evaluating oxygen transfer data as long as the effective average dissolved oxygen concentration in the aeration zone is equal to the effective average dissolved oxygen concentration in the accumulation zone. Baillod [1] has further shown that the variation in saturation concentration, C*, leads to an "apparent" $K_{La}-K_{La}'$,—which may be 10 to 30% less than the true K_{La} as expressed in Equation 5.2. For purposes of modeling and data analysis, Equation 5.2 may be rewritten as:

$$\frac{dC}{dt} = K_{La}' (C_\infty^* - C) \qquad (5.10)$$

where C_∞^* = average dissolved oxygen saturation concentration attained after aerating a nonoxygen-demanding solution for an extended period of time, mg/L

Note that C_∞^* is a measurable quantity, whereas C* is extremely difficult to measure and varies during the course of an unsteady-state reaeration test as the mean oxygen content of gas bubbles increases as C_∞^* is approached. The use of C_∞^* and K_{La}' facilitates modeling and data analysis of aeration systems and forms the basis for most current oxygen transfer models.

RECOMMENDED STANDARD METHOD
FOR MODELING AND ANALYZING
UNSTEADY-STATE TEST DATA [1]

The basic model recommended for the analysis of both surface and subsurface clean water, unsteady-state oxygen transfer test data is:

$$\frac{dC}{dt} = K_{La}' (C_\infty^* - C) \quad \text{(differential form)} \qquad (5.11)$$

which on integration becomes:

$$\ln \frac{C_\infty^* - C}{C^* - C_o} = K_{La}'t \quad \text{(logarithmic form)} \qquad (5.12)$$

or

$$C = C_\infty^* - (C_\infty^* - C_o) \exp(-K_{La}'t) \quad \text{(exponential form)} \qquad (5.13)$$

where C_∞^* = average dissolved oxygen saturation concentration attained at infinite time, M/L^3

K_{La}' = apparent volumetric mass transfer coefficient, T^{-1}

C_o = dissolved oxygen concentration at t = 0, estimated from the model, M/L^3

C = effective average dissolved oxygen concentration in the liquid phase, M/L^3

t = time

and the symbols, M, L, T and F are employed to denote dimensions of mass, length, time and force, respectively.

This model applies to a given aeration system in a given tank under steady-state hydraulic conditions. Oxygen transfer need not occur throughout the entire tank volume and the tank need not be completely mixed. However, in the case of localized transfer, the model assumes that the effective average dissolved oxygen concentration and saturation value of the transfer zone are equal to that of the entire tank. In any case, the effective average dissolved oxygen concentration in the tank, C, represents a bulk average:

$$C = \frac{\int \hat{C} dV}{V} \tag{5.14}$$

If the results of multiple sample points are to be analyzed by a simple average, the sample locations should be chosen so that each samples an equal portion of the tank volume.

In the application of this model to unsteady-state test data, it is recommended that:

1. The exponential form should be employed and fit to the test data by nonlinear regression.

2. Values of the parameters K_{La}', C_∞^* and C_o should be estimated from the model. Individual measured values of C_∞^* and C_o are not to be used as model parameters. Likewise, tabulated book values of dissolved oxygen surface saturation concentrations are not to be used to estimate C_∞^*.

3. An effort should be made to gather valid dissolved oxygen data over as wide a range as possible. Truncation of data at values of C less than 20% of C_∞^* is allowable to avoid lingering effects of the deoxygenation technique. However, good data at low dissolved oxygen values should be used in the estimation because they are important for assessing model adequacy. A guideline which may be applied for more precise determination of the low truncation point when sampling intervals are small is to locate an inflection point by differentiating the data and truncate data at C values below 1.5 times the C at the inflection point. However, in no case should values of C greater than 30% of C_∞^* be truncated. Truncation of data at dissolved oxygen values approaching C_∞^* is strongly discouraged and is recommended only when the alternative "best fit log deficit" method is employed.

4. When lack of a computing facility prevents the application of nonlinear regression to the exponential form of the model, a linear regression applied to the logarithmic form of the model is an acceptable alternative technique. When this technique, termed the best fit log deficit, is used, the value of C_∞^* is to be estimated from the model as the value giving the minimum residual sum of squares on the log deficit plot.

5. In most cases, calculation of standard clean water and dirty water, respiring system oxygen transfer rates [standard oxygen requirement (SOR) and actual oxygen requirement (AOR)] should be based on the estimated values of K_{La}' and C_∞^* along with given or separately determined values of α', θ' and β. This approach is relatively straightforward and simple. However, it is acceptable to base the calculations of SOR and AOR on a true volumetric mass transfer coefficient, K_{La}'. In this case, values of α' and θ' based on ratios of K_{La}' values should be distinguished from values of α and θ based on K_{La} values.

DETERMINATION OF OXYGEN REQUIREMENT

The first step in sizing an aeration system is to determine the oxygen demand which will be exerted on the aeration system. An approach to determining the oxygen demand in oxidation ditch systems is presented in Chapter 9.

Empirical approximations to oxygen demand are possible. Factors can be applied which relate the pounds of oxygen required per pound of BOD_5 applied or removed. These factors or ratios are functions of the particular biological treatment system considered. Generally, the longer the retention time (solids or liquid), the greater the amount of oxygen required to remove a pound of BOD_5. This phenomenon is basically a result of the BOD_5 descriptor (the BOD test does not measure a precise quantity or oxygen requirement, and is dependent on incubation time and temperature, which may or may not relate to actual treatment conditions), and the extent of autooxidation of biological solids present in the system. In general, the longer the solids retention time, the greater the amount of autooxidation and the larger the oxygen requirement for cell metabolism. For example, two extremes in the oxygen requirement ratio would be a high-rate activated sludge process which might operate with a 0.6 or 0.7 ratio, and an extended aeration activated sludge system which might operate with a ratio as high as 2.0. Table 5.1 summarizes parameters for various activated sludge processes, along with the ratios of pounds of oxygen required to remove a pound of BOD_5 and suggested design values. Obviously, the selection of the O_2/BOD_5 ratio greatly influences equipment and energy requirements. The effects of nitrification on oxygen requirements must be considered. These effects are discussed in Chapters 4 and 9.

Once the process oxygen demand is calculated, a conversion must be made to transfer process oxygen demand, AOR, into standard demand, SOR. Standard oxygen requirement is derived from standardized test data taken from test facilities and field installations. The AOR may be converted to SOR by use of Equation 5.15:

$$\text{SOR} = \cfrac{\text{AOR}}{\alpha \cfrac{(\beta C_{\infty}^{*} - C_L)}{C_s} \theta^{T-20}} \tag{5.15}$$

where C_L = desired aeration basin DO, mg/L
C_s = saturation DO at sea level and 20°C in clean water, mg/L
α = relative rate of oxygen transfer in wastewater to that in clean water (K_{La} wastewater/K_{La} clean water)
β = ratio of saturation values for oxygen in wastewater to that in clean water
θ = temperature correction factor, usually 1.024

The values for α and β must be chosen with some care. Alpha has been shown to vary from 0.3 for low-turbulence aerators in poorly treated wastes to greater than 1.0 for well treated wastes and high-turbulence aerators. Generally, values range from 0.5 to 0.95 for municipal wastes, and may be greatly affected by industrial wastes. Beta values usually range from 0.90 to 0.97. Industrial wastes can severely depress beta values. Tables 5.2 and 5.3 present the data required for the use of Equation 5.15.

The value of C_{∞}^{*} may be calculated by a number of methods for submerged aerators. For surface aerators C_{∞}^{*} is usually assumed to be the surface saturation value in clean water at wastewater temperature (Table 5.2). For submerged aerators C_{∞}^{*} may be estimated from:

$$C_{\infty}^{*} = H\,Y_d \quad Pa + \left(\frac{\gamma\,Z_d}{2 \times 144} \right) \tag{5.16}$$

where H = Henry's law constant, mg/L/psi
Y_d = mol fraction of oxygen in air at bubble release, 0.209 for air
Pa = atmospheric pressure, psia
γ = weight of water, 62.4 lb/ft^3
Z_d = aerator submergence, ft

A demonstration of the calculation of SOR for three different aeration systems is presented as Example 5.1. In Example 5.1 the effect of aerator type, due to both submergence and alpha effects, may be seen.

Table 5.1. Parameters for Various Activated Sludge Processes

Parameters	Process					
	Conventional	Tapered Aeration	Step Aeration	Contact Stabilization	High-Rate Activated Sludge	Extended Aeration
F/M Ratio	0.2-0.5	0.2-0.5	0.2-0.5	0.2-0.5	2.0-3.5	0.05-0.2
Detention Time V/Q, hr	4	4	4-6	0.2-1.5 contact (3-4 hr) total	0.5-3	24
MLSS, mg/L	1000-3000	1000-3000	2000-3000	2500 contact (4000-8000) reaeration	4000-10,000	3500-5000
Approximate % Volatile Solids	65-75	70	70	70	70-80	45-50
SRT, days	6-15	6-15	6-15	6-15	<3	20-30
Return Sludge Ratio R/Q	0.1-0.3	0.1-0.3	0.2-0.35	0.4-1.25	0.5-4	0.5-1.5
Organic Loading, lb $BOD_5/10^3$ ft^3/day	30-40	30-40	40-60	60-75	100	15-25
Overall Efficiency, Percent BOD_5 Reduction	90-95	90-95	90-95	85-90	50-75	90-98
lb Oxygen Required to Remove 1 lb of BOD_5, L	1.0-1.2	1.0-1.2	1.0-1.2	0.8-1.2	0.5-1.0	1.2-2.0
Suggested Design Value	1.1	1.1	1.1	1.0	0.8	1.4

Table 5.2. Saturation Dissolved Oxygen in Clean Water at One Atmosphere Pressure [7]

Temperature, °C	Saturation Dissolved Oxygen Concentration, mg/L
5	12.75
6	12.43
7	12.12
8	11.83
9	11.55
10	11.27
11	11.01
12	10.76
13	10.52
14	10.29
15	10.07
16	9.85
17	9.65
18	9.45
19	9.26
20	9.07
21	8.90
22	8.72
23	8.56
24	8.40
25	8.24
26	8.09
27	7.95
28	7.81
29	7.67
30	7.54

Table 5.3. Atmospheric Pressure Corresponding to Altitude

Elevation, ft	mm Hg	Elevation, ft	mm Hg	Elevation, ft	mm Hg
Sea Level	760.0	2500	693.7	5000	632.2
500	746.3	3000	681.0	5500	620.5
1000	733.1	3500	668.6	6000	609.1
1500	719.6	4000	656.4	6500	597.7
2000	706.7	4500	644.2	7000	586.5

Example 5.1. Calculate the standard oxygenation requirement for a surface rotor ($\alpha = 0.90$), a jet aerator ($\alpha = 0.90$) and fine bubble diffuser ($\alpha = 0.50$) for a 10-foot-deep oxidation channel under the following conditions:

- AOR   = 100 kg/hr
- T   = 25°C
- β = 0.95
- C_L = 2.0 mg/L
- plant elevation = 5000 ft

For the surface rotor:

From Table 5.2: Saturation DO at sea level and 25°C = 8.24 mg/L
From Table 5.3: Atmospheric pressure = 632.2 mm Hg

Thus:

$$C^*_\infty = \frac{632.2}{760.0} \times 8.24 = 6.85 \text{ mg/L}$$

Then:

$$SOR = \frac{100 \times 9.07}{0.90(0.95 \times 6.85 - 2) \, 1.024^{25 - 20}}$$

$$SOR = 198.6 \text{ kg/hr}$$

For jet aerators: Assume aerator submergence is 8 ft

First calculate C^*_∞:

$$C^*_\infty = 2.71 \times 0.209 \quad 14.7 \quad \times \left(\frac{632.2}{760.0} + \frac{62.4 \times 8}{144 \times 2} \right)$$

$$C^*_\infty = 7.91 \text{ mg/l}$$

Then:

$$SOR = \frac{100 \times 9.07}{0.90(0.95 \times 7.91 - 2)1.024^{25 - 20}}$$

$$SOR = 162.3 \text{ kg/hr}$$

For fine bubble aerators: Assume 10-ft aerator submergence

$$C^*_\infty = 2.71 \times 0.209 \quad 14.7 \times \frac{632.2}{760.0} + \frac{62.4 \times 10}{144 \times 2}$$

$$C^*_\infty = 8.15 \text{ mg/L}$$

$$SOR = \frac{100 \times 9.07}{0.50 (0.95 \times 8.15 - 2)1.024^{25-30}}$$

$$SOR = 280.5 \text{ kg/hr}$$

REFERENCES

1. Baillod, C. R., and L. C. Brown, "Standard Method for the Evaluation of Oxygen Transfer by the Clean Water Unsteady-State Method," Unpublished Report, Am. Soc. for Civil Eng., Oxygen Transfer Task Force (March, 1980).
2. Baillod, C. R., "Review of Oxygen Transfer Model Requirements and Data Interpretation," in *Proceedings of the Workshop: Toward an Oxygen Transfer Standard*, U.S. EPA, 600/9-78-021:17-27 (April, 1979).
3. Downing, A. A., and A. G. Boon, "Oxygen Transfer in the Activated Sludge Process," *Advances in Biological Waste Treatment*, ed. by W. W. Eckenfelder and B. J. McCabe (New York, 1963), p. 131.
4. Ewing, L., D. T. Redmon and J. D. Wren, "Testing and Data Analysis of Diffused Aeration Equipment," *J. Water Poll. Control Fed.*, 51:2384 (1979).
5. Redmon, D. T., M. Mandt and J. D. Wren, "Oxidation Transfer Data Interpretation: Non-Steady State Clean Water Tests," in *Proceedings of the Workshop: Toward an Oxygen Transfer Standard*, U.S. EPA 600/9-78-021 (April, 1979).
6. Mandt, M. G., "Improvements in Aeration Testing and Performance Evaluation," presented at 50th Annual WPCF Conference, Philadelphia, PA (1977).
7. *Standard Methods for the Examination of Water and Wastewater*, 15th Edition (New York: APHA, 1981).

CHAPTER 6

MIXING AND HYDRAULICS

MIXING CONSIDERATIONS

Continuous loop reactors (CLR) present a unique combination of mixing capabilities. Mixing considerations of interest in a CLR are:

1. oxygen and substrate dispersion and contact with microorganisms
2. off-bottom solids transport
3. control of concentration gradients affecting reactor kinetics
4. control of short circuiting
5. macromixing and micromixing effects on kinetics and flocculation

In a CLR the wastewater feed system and return activated sludge are normally introduced into the CLR at a point. The mixed liquor contents of the CLR continuously move past this point, in effect allowing the metering of feed stream and return sludge into the moving mixed liquor stream, providing dilution and shock loading protection. In conventional complete mix systems, mixing and rapid turnover of the complete reactor contents are necessary to transport substrate to the far reaches of the reactor, away from the feed point or points. In a plug flow reactor the feed stream concentrates at the inlet end. Shock or toxic loads are magnified.

Oxygen is transferred in an aeration zone, persists in a nonaerated aerobic, accumulation zone, and finally becomes limited in the anoxic zone (see Figure 6.1). Since it is possible to arrange for considerable channel length and time between aeration zones, CLRs are eminently suited for nitrification and denitrification. The positive control of anoxic mixed liquor into the aeration device provides a favorable oxygen driving gradient for oxygen transfer.

In a CLR, conservation of momentum is employed. Once the channel contents are accelerated to channel velocity, the hydraulic power to maintain circulation is only that required to overcome friction and bend loss. The

85

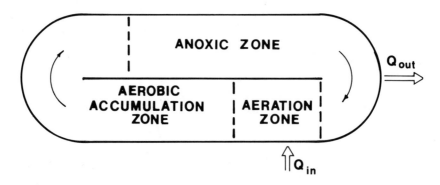

Figure 6.1. Oxygen zones of a CLR.

circulation or convective mixing employed reinforces itself, as opposed to dissipation through random eddies. As a result, it is possible to maintain velocity for solids suspension at much lower volume-specific power densities than in alternative systems. In addition, heavy solids can be transported at, or off, the bottom of the channel until aeration zones are reached. The high energy associated with an aeration zone resuspends solids that may tend to settle in nonaerated zones. Continuity considerations further dictate that once channel velocity is reached, the mean horizontal velocity is identical throughout the channel. The secondary turbulence created by mean channel velocity is responsible for channel mixing and solids suspension.

Oxygen concentration, organic carbon, ammonia and nitrate concentration gradients affect CLR kinetics. A properly functioning CLR normally operates at very low concentration gradients due to feed stream dilution and the mixing provided. As previously stated, a CLR can be considered a completely mixed reactor from a macromixing viewpoint; i.e., complete mix models can be used to describe macromixing characteristics. It is, however, the small concentration gradients within a CLR that provide much of its versatility.

It is normal and good design practice to arrange for feed streams to the CLR to enter at or just upstream of the aeration zone. Good local mixing and dispersion take place through the aeration zone. The well mixed contents of the CLR then continue around the CLR with effluent withdrawal normally accomplished just upstream of the feed stream point. In this fashion the feed stream must make at least one complete revolution of the CLR before any possibility of short-circuiting in the effluent. In some respects this is similar to plug flow systems. It could be said that over a short term (one revolution) a CLR exhibits some plug flow characteristics and over long term (many revolutions) a CLR exhibits some complete mix characteristics. In any event, the CLR uniquely combines the better attributes of both in providing an effective treatment system.

Often overlooked is the effect of mixing on flocculation of solids in bioreactors. Engande [1] and Mandt [2] have discussed the effect of turbulence on flocculation and solids settleability. In a CLR there are two zones of mixing (Figure 6.2): a high-energy zone associated with the aeration or mixing device, and a low-energy zone associated only with channel circulation. A transition zone between the two might also be considered as energy dissipates from a high to a low level.

The high-energy zone will generally have a mean velocity gradient, $\bar{G}$, in excess of 100 $\sec^{-1}$. $\bar{G}$ may be defined as:

$$\bar{G} = \sqrt{\frac{P}{\mu V}} \tag{6.1}$$

where   P = power transferred to the liquid, ft-lb/sec
        μ = dynamic viscosity of the liquid, lb-sec/ft^2
        V = volume of the aeration zone, ft^3

The mean velocity gradient of the nonaerated zone of a CLR is normally less than 30 $\sec^{-1}$. Low $\bar{G}$ values in activated sludge systems allow for bioflocculation of mixed liquor solids [1,2]. Thus, the nonaerated portion of a CLR provides conditions conducive to flocculation.

The unique mixing capabilities of CLRs may explain to a large degree their superior treatment capabilities as compared to alternative biological treatment systems as reported by the U.S. Environmental Protection Agency (EPA) [3]. The effect of mixing on organic carbon, ammonia, nitrate, and solids removal should not be underestimated.

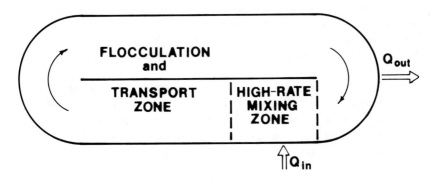

Figure 6.2. Mixing zones of a CLR.

HYDRAULIC CONSIDERATIONS

Continuous loop reactors can be analyzed by use of open channel hydraulics. The frictional losses, h_f, in the channel can be estimated from Manning's equation:

$$h_f = \frac{\eta^2 L V^2}{(1.49)^2 R^{4/3}} \qquad (6.2)$$

where η = Manning's number
        L = mean channel length, ft
        V = mean channel velocity, ft/sec
        R = hydraulic radius

The hydraulic radius is equal to the channel cross-sectional area divided by the wetted perimeter. (See Example 6.1.)

Bend losses are more difficult to estimate. Bend losses, h_b, can be estimated from the general head loss equation expressed as:

$$h_b = f_c \frac{V^2}{2g} \qquad (6.3)$$

where f_c = coefficient of curve resistance
        V = mean channel velocity, ft
        g = gravitational constant, ft/sec^2

The coefficient of curve resistance is a function of Reynolds number, the ratio of bend radius to channel width, the ratio of channel depth to channel width, and the degree of curvature. Shukry [4] has published data for coefficient of curve resistance for vertical wall channels. (See Figure 6.3) Though no sound quantitative data for trapezoidal channels presently exist, Shukry's data can be applied using maximum channel width and an appropriate safety factor. Vertical wall channels will have less curve resistance than trapezoidal channels. Thin channels and channels with gradual bends will have the lowest coefficient of curve resistance.

Additional channel losses result from back-pumping by the aeration device and momentum transfer losses by the aeration or propulsion device. Mixing capabilities of various types of CLRs are discussed in Chapter 7.

The momentum flux from all propulsion devices within a CLR creates a small head, y, which can be expressed as:

$$y = \frac{\Sigma M_f}{g A \gamma_w} \qquad (6.4)$$

where M_f = momentum flux from each device
$\quad\quad$ A $\;$ = channel cross-sectional area, ft^2
$\quad\quad$ γ_w = mass of water, lb/ft^3

Equating losses to available head yields the mean channel velocity estimate:

$$y = \left(\frac{n^2 L}{(1.49)^2 R^{4/3}}\right) + \frac{f_c}{2g} V^2 \tag{6.5}$$

Example 6.1. Calculate the head available in a CLR containing two surface rotors each capable of delivering 1000 lb of thrust to the channel liquid when the CLR is 30 ft wide and 10 ft deep.

$$y = \frac{\Sigma M_f}{g \, A \, \gamma_w} = \frac{2(1000)}{32(30)\,(10)\,(62.4)} = 0.00334 \text{ ft} \tag{6.6}$$

In many CLRs bend losses represent 90% or more of channel losses. Good bend design is of paramount concern where channel velocity is critical. Many CLRs have one or more turning baffles in each bend in order to reduce bend losses. For bends with multiple channels, each channel can be analyzed individually and summed according to Equation 6.7.

$$\frac{1}{h_b} = \frac{1}{h_{b_1}} + \frac{1}{h_{b_2}} + \ldots + \frac{1}{h_{b_n}} \tag{6.7}$$

where h_{b_1} through h_{b_n} = the head loss for each individual channel within a bend divided by n – 1 turning baffles

The summation is analogous to electrical circuits in parallel.

Example 6.2. An individual channel bend has a coefficient of curve resistance, f_c = 1. When divided by turn baffles, the resulting individual channels have f_{c_1} = 1.5, f_{c_2} = 0.7, and f_{c_3} = 0.5. What is the reduction in head loss through the bend dividing the channel?

$$\frac{1}{f_c} = \frac{1}{f_{c_1}} + \frac{1}{f_{c_2}} + \frac{1}{f_{c_3}} \tag{6.8}$$

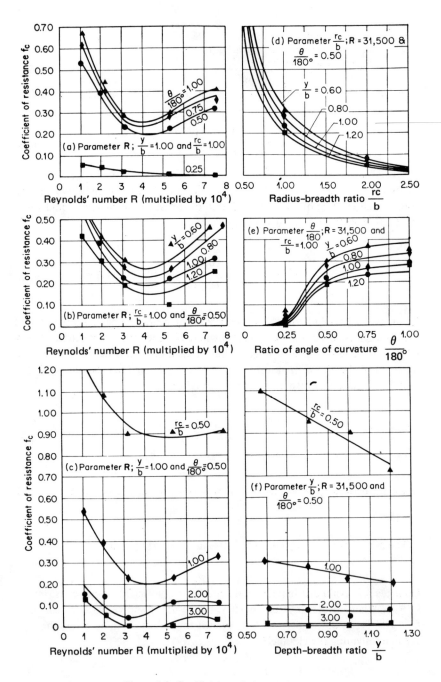

Figure 6.3. Coefficients of curve resistance.

$$\frac{1}{f_c} = \frac{1}{1.5} + \frac{1}{0.7} + \frac{1}{0.5} = 4.1 \tag{6.9}$$

$$f_c = 0.244$$

The bend loss is thus reduced to 24.4% of the undivided bend loss. The dramatic effect of turning baffles is obvious. If the turning baffles are more than thin-walled structures, the additional loss due to flow impediment must be considered. For thin-walled baffles drag and frictional losses are normally negligible.

Note that the inner channel will always have the highest coefficient of curve resistance due primarily to the sharper radius of curvature. Turning baffles are often offset in the bend attempting to equalize flow between bend channels.

REFERENCES

1. Engande, A. J., et al., "Oxygen Concentration and Turbulence as Parameters of Activated Sludge Scale-Up," presented at Water Resources Symposium No. 6, Center for Research in Water Resources, University of Texas at Austin (1973).
2. Mandt, M. G., "Multi-Stage System for Wastewater Oxidation," U.S. Patent No. 4,206,047.
3. Ettlich, W. F., "A Comparison of Oxidation Ditch Plants to Competing Processes for Secondary and Advanced Treatment of Municipal Wastes," U.S. EPA Report 600/2-78-051 (March, 1978).
4. Shukry, A., "Flow Around Bends in an Open Flume," Transactions of the American Society of Civil Engineers, Paper No. 2411.

CHAPTER 7

PROPRIETARY PROCESSES

Since the introduction of the original Pasveer oxidation ditch, a variety of process and equipment modifications have evolved. Many of these are based on harnessing the horizontal component of flow from a particular aeration device. Most of these modifications are proprietary to one degree or another, based on a particular area of expertise of the introducing individual or company or on certain patented features.

This chapter presents features of and design procedures for the various process and equipment modifications presently employed in continuous loop reactors. The authors have endeavored to present sufficient factual information on each process to allow the engineer to carry out or verify a complete design. It is recommended, however, that the process originator be consulted during the design and application of each process.

Continuous looped reactor configurations and aeration devices can be grouped into six broad classifications:

1. Carrousels and activox plants using vertical surface aerators.
2. Jet aeration channels using jet aerators.
3. Orbal plants using rotor-mounted perforated disks in a series of loops.
4. Single loop plants using brush or cage aerators.
5. Barrier ditches using submerged or surface turbines and draft tubes.
6. Combined systems using both diffused aeration and a propulsion device.

Carrousels are offered by Dwars, Heederick and Verhay, Ltd., Envirotech and Activox. Jet aeration channels are offered by Pentech Houdaille and Fluidyne Corporation. Orbal systems are offered by Envirex, single loop plants by Lakeside, Passavant and Cherne. The draft tube channel barrier ditch is offered by Mixing Equipment Company. Scriebner offers a rotating diffuser CLR, and Fluidyne offers a CLR with fine bubble diffusers and auxiliary jets or propellers for propulsion.

CARROUSEL

The Carrousel activated sludge system was developed by Dwars, Heederik, and Verhay, Ltd., consulting engineers in Holland in the late 1960s [1]. Their goal was to retain some of.the advantages of the commonly used brush aerator oxidation ditch system, but to develop a more energy-efficient and lower-cost system using a deeper aeration basin. Since this initial work, over 100 Carrousel activated sludge systems ranging from 0.5 MGD to 300 MGD have been put into operation. Envirotech Corporation has licensed the Carrousel system for use in the United States and Canada. A schematic of the Carrousel system is shown in Figure 7.1. It consists of vertically mounted low-speed surface aerators in an aeration basin, with partitions used to establish a continuous channel.

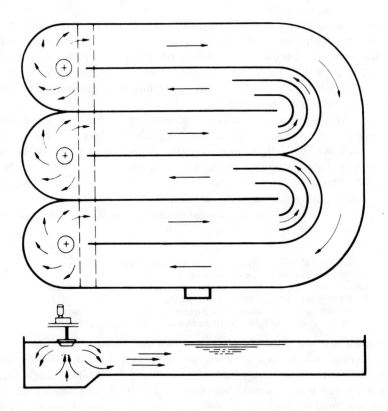

Figure 7.1. Schematic of Carrousel system.

Wastewater and return sludge are mixed in the first aeration zone. A flow velocity of 1 ft/sec is maintained in the channel due to the pumping action of the aerators. This pumping action is achieved by lining up the partition with the aerator in such a manner that the aerator pushes the flow through the aeration zone and out into the channel. The flow passes through successive aeration zones and around a final outer loop where the effluent exits over a weir just prior to the first aeration zone.

Proper basin geometry and aerator design consideration are necessary to provide proper channel velocities. The key design factors are aerator impeller diameter, aerator rotational speed, aerator impeller submergence and design, position of the center partition, channel depth, channel width, aerator layout and aeration basin volume. The effect of all these factors has been integrated into a Carrousel hydraulic design model to assure proper channel velocities.

The Carrousel has the same basic capabilities common to all continuous loop reactors:

1. complete mix capability
2. channel plug flow
3. aerator application methods
4. conservation of momentum through convective mixing, resulting in the ability to suspend solids at low power densitiies

The basis for complete mix operation is that at a 1-ft/sec channel velocity, the volume of liquid circulating through the channels may be between 30 and 50 times the influent flow. While an amount equal of the influent flow is continually displaced over the effluent weir, tremendous dilution is provided as the influent is combined with the mixed liquor recirculating through the channels.

The liquid in the channels completes the circuit every 5 to 20 min, depending on the channel length and design loading. Thus, the flow pattern prevents short-circuiting and still offers the buffering features of a complete mix system.

In certain sections of a CLR, some of the benefits of the plug flow mode of operation can be realized. One of these is the gentle mixed liquor solids reflocculation period after the aeration zone, which can result in improved sludge flocculation characteristics and improved settling and clarification in the final clarifier.

Another plug flow feature is that the mixed liquor dissolved oxygen level is reduced to zero at some point. This then promotes denitrification (utilization of the nitrate by the bacteria in place of oxygen). The amount of denitrification that occurs depends on the endogenous respiration rate of the mixed liquor solids and the length of the anaerobic channel section. These factors can be taken into account in design to provide desired levels of denitrification. Benefits associated with this include nitrogen removal and reuse of the nitrate oxygen to satisfy a portion of the oxygen demand. Thus, for a given sludge age or level of sludge stabilization, the net oxygen requirements for a CLR can be reduced by 10 to 25%, resulting in lower power requirements. The denitrification process also restores some of the alkalinity which is depleted during the nitrification process, reducing the amount of chemicals required to control the pH for nitrification.

The aerator application method results in a number of advantages, including efficient oxygen transfer rates and a very efficient means of mixing the activated sludge basin. Only a portion of the aerator horsepower is generally used to generate channel flow, so that Carrousel aerators are normally designed on the basis of oxygenation requirements [2]. To produce a nitrified or extended aeration effluent, long activated sludge aeration times are normally required, resulting in a mixing limited design for conventional activated sludge systems. The CLR will not require the same level of mixing horsepower, and thus significant energy savings are achieved in the production of a high-quality effluent.

With all of the aerators located at one end of the Carrousel aeration basin, a very high power intensity occurs in the aeration zone (roughly 4-8 hp/1000 ft^3, as compared to 0.75-1.0 hp/1000 ft^3 for conventional aeration). Figure 7.2 shows that the higher power intensity improves the aerator oxygen

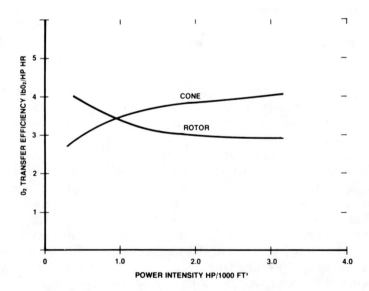

Figure 7.2 Transfer efficiency vs power intensity for surface aerators [3].

transfer efficiency [3]. Thus, regardless of mixing, Carrousel aerators are applied in a very efficient operating mode.

Convective mixing capability in CLRs can also result in a significant power savings during low loading periods. Figure 7.3 shows results from a two-aerator system at Osterwoolde. This type of mixing and flow configuration makes it possible to use fewer aerators, reducing capital costs and maintenance requirements. It is often advisable to cover the aeration zones of Carrousel aerators to prevent misting and icing problems.

Carrousel Treatment Performance

In the Netherlands, the overall wastewater treatment objective is normally to reduce the total biochemical oxygen demand for both total carbonaceous and nitrogenous demand to a minimum value. An effluent tax is imposed on discharges based on the quantity of total oxygen demand in the effluent. Because of this treatment requirement, Carrousel systems there normally have been operated in an extended aeration mode. Extended aeration produces an extremely stable sludge, and because of this the Carrousel waste sludge is suitable for direct land disposal for agricultural purposes.

These extended aeration systems have normally been designed for sludge age (SRT) values in excess of 25 to 30 days, with aeration detention times in the range of 18 to 28 hr. Tables 7.1 and 7.2 show typical results for such a Carrousel system design at Lichtenvoorde, Holland. The results show that the effluent BOD_5 values were normally less than 10 mg/L and high levels of

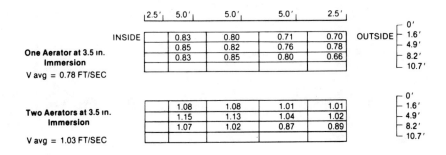

Figure 7.3. Channel velocity profiles from a Carrousel plant.

Table 7.1. Lichtenvoorde Carrousel Operating Results

Sample Date[a]	Influent				Effluent						
	pH	BOD₅, mg/l	COD, mg/l	TKN, mg/l	pH	BOD₅, mg/l	COD, mg/l	NH₃-N, mg/l	NO₂-N, mg/l	NO₃-N, mg/l	TKN, mg/l
6/12/72	7.6	330	705	58.7	7.4	4	58	1.5	0.04	26	3.2
6/20	—	—	1760	186	7.3	7	71	1.4	0.12	47	4.5
6/28	8.6	400	1010	92.6	7.7	3	56	1.2	0.05	36	2.5
7/20	7.8	210	480	26.3	7.3	8	52	1.1	0.04	27	2.7
7/25	8.5	520	1170	117	7.5	2	39	0.9	0.12	28	2.4
8/9	7.3	300	615	37.6	7.5	2	31	0.6	0.02	11	2.1
8/24	8.6	610	1270	187	7.5	1	31	0.6	0.34	10	2.5
9/5	8.3	710	1000	167	7.9	2	43	0.6	—	35	2.5
9/21	7.8	12	47	40	7.3	3	29	0.5	0.04	33	1.5
10/4	8.5	815	1540	157	7.6	3	50	0.4	0.04	36	2.4
10/17	8.5	1270	1990	184	7.2	4	41	0.7	0.03	80	2.9
11/1	7.3	980	1540	133	7.0	3	45	0.7	0.06	83	3.2
11/14	7.3	188	415	32	7.2	5	52	0.6	0.02	30	1.3
11/28	8.0	720	1370	112	6.7	6	47	0.7	0.03	68	2.6
12/13	8.8	390	1100	97	6.9	4	57	0.2	0.02	60	2.4
1/4/73	6.5	1040	2070	161	7.1	12	108	3.9	0.08	63	11
1/13	8.7	1000	1600	153	7.9	9	81	0.6	0.03	56	4.1
2/1	8.5	650	1330	91	7.2	9	82	0.2	0.01	39	4.3
2/13	8.9	850	1310	87	7.3	8	61	0.9	0.02	13	4.3
3/13	9.2	680	1390	128	6.9	4	85	0.2	0.02	57	2.7
3/23	8.0	460	920	82	6.9	4	62	1.1	0.06	87	3.9
4/11	7.5	580	1050	104	7.2	5	56	0.5	0.03	50	2.5
4/25	7.6	710	1870	108	7.1	3	69	0.3	0.05	58	3.7
5/4	7.6	1660	4040	29	7.0	3	37	0.2	0.25	35	1.7
5/22	9.5	950	1700	123	7.5	7	52	0.4	0.08	35	1.5
6/3	8.0	740	1880	103	7.2	3	34	0.1	0.06	29	2.2
Average	8.1	663	1315	108.3	7.3	5	55	0.8	0.06	44	3.1

[a] 9-hr composite samples.

Table 7.2. Summary of Operation Results and Removal Efficiencies
for the Lichtenvoorde Treatment Plant

	Influent	Effluent	% Removal
BOD_5, mg/l	663	4.6	99.3
COD, mg/l	1316	54.9	95.8
Kjeldahl (TKN)	108.3	3.1	97.1
Ammonia (NH_3-N)		0.8	
Nitrite (NO_2-N)		0.06	
Nitrate (NO_3-N)		43.5	
Degree of Nitrification			93.2
Nitrogen Removal[a]			56.9

[a]Lichtenvoorde plant was not designed nor operated for nitrogen control.

nitrification were obtained. The system also resulted in about 57% nitrogen removal by utilizing the high level of nitrate produced from nitrification. A summary of effluent BOD data taken from four Carrousel wastewater treatment plants in the Netherlands is shown in Figure 7.4. The mean effluent BOD_5 value was 3.7 mg/L, and 90% of the time the effluent BOD_5 was less than 7.5 mg/L. For the same plants, the mean effluent ammonia-nitrogen value was less than 1 mg/L, as shown in Figure 7.5.

Design Considerations and Procedures

Even though the Carrousel has normally been used in an extended aeration mode, the process is not contingent on any particular loading rate. It may be used in the conventional activated sludge process preceded by a primary settling basin as well as in nitrification only systems, or it can be used in extended aeration applications. The first step in the design approach is to determine the level of treatment required, which will then fix the food mass ratio or SRT required for the Carrousel system. If BOD removal only is required then a sludge age of approximately 5 days would be used for design. A BOD removal only application would not take full advantage of the Carrousel mixing energy savings but could still take advantage of its other aerator application features. For nitrification designs, the sludge age would be in the range of 10 to 20 days, depending on the wastewater temperature. This is based on providing a 2:1 safety factor and using Downing's equation for predicting the minimum nitrification sludge age as a function of temperature [4]. For an extended aeration design, a Carrousel system would be designed on the basis of about a 30-day sludge age or 20- to 24-hr hydraulic detention time to achieve complete sludge stabilization as well as nitrification and low effluent

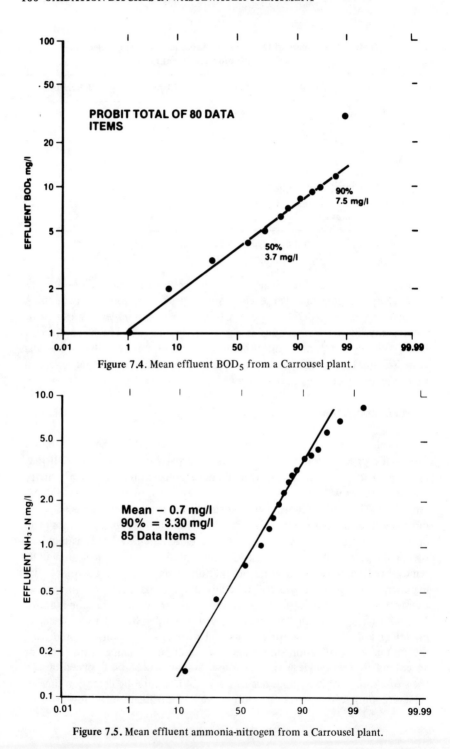

Figure 7.4. Mean effluent BOD$_5$ from a Carrousel plant.

Figure 7.5. Mean effluent ammonia-nitrogen from a Carrousel plant.

BOD$_5$ values. Once the sludge age is selected, the following equation is then used to determine the necessary aerobic volume in the Carrousel system [5]:

$$V = \frac{(SRT)\,(Yn)\,(\Delta BOD)}{MLSS\,(8.34)} \qquad (7.1)$$

where V   =   aeration volume, MG
       BOD =   lb of BOD removed per day
       Yn   =   Net solids production coefficient, lb SS/lb BOD$_R$
       MLSS =   mixed liquor suspended solids concentration, mg/L

The MLSS concentration is selected at approximately 3500 mg/L to provide conservative solids floor loadings in the secondary clarifier. However, it has been common to operate Carrousel systems in Europe at mixed liquor levels of 4000 to 7000 mg/L.

One of the key design parameters in Equation 7.1 is the net solids yield coefficient. A general set of curves has been developed for taking into account the synthesis yield coefficient, the endogenous respiration rate coefficient, and the effect of SRT and influent inert solids. Figure 7.6 summarizes a typical range of values for Yn.

Inert solids are defined as solids entering the wastewater treatment system that are not biodegraded. These consist of inorganic solids and nonbiodegradable volatile solids. The quantity of inert solids may range from 30 to 100 mg/L depending on the degree of primary treatment. The higher the level of

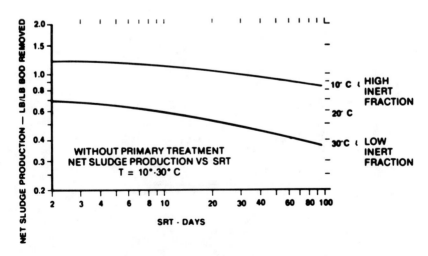

Figure 7.6. Net sludge production from a CLR.

inert solids, the greater the net sludge yield and the greater the aeration tank volume necessary for a given sludge age.

For a given amount of denitrification, the volume in Equation 7.1 must be increased to account for the denitrification anoxic volume. The denitrification anoxic volume is determined by Equation 7.2:

$$VD = \frac{N}{DNR \ (MLSS) \ 8.34} \tag{7.2}$$

where VD = denitrification volume, MG
      N = quantity of nitrate nitrogen to reduce, lb/day
    DNR = specific solids denitrification rate, lb N/lb MLSS-day

The next step in the design is to determine the aerator horsepower requirements. This is done by determining the quantity of oxygen necessary for the level of treatment desired. Figure 7.7 is an example of a curve used to select the amount of oxygen for BOD removal as a function of the design sludge age. In addition to this oxygen requirement, the amount of oxygen necessary for nitrification must be added. This is normally taken as 4.5 lb of oxygen per pound of ammonia-nitrogen oxidized. The total quantity of oxygen required can then be reduced by taking into account the oxygen made available to the system through denitrification. This quantity is 2.6 lb of oxygen per pound of nitrate-nitrogen reduced. Once the oxygen requirements are determined, the aerator standard oxygen transfer efficiency value converted to a mixed liquor condition is used to determine the total horsepower for the Carrousel system. Once the total horsepower is selected, the Carrousel basin layout can be evaluated. The number of aerators selected will determine the

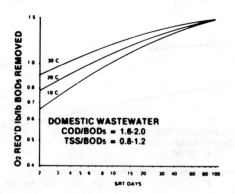

Figure 7.7. Oxygen requirements for carbonaceous BOD_5 removal vs SRT and temperature (add O_2 for nitrification as required).

aerator size. These aerators can be laid out in a variety of configurations (Figure 7.8). Normally, the channel depth is in the range of 1.2 times the impeller diameter and the channel width is twice the depth. The hydraulic model developed by Dwars, Heederik, and Verhay, Ltd., is used to optimize the aeration basin and aerator design to provide maximum channel velocity. The design selected can be varied depending on a variety of operating and site location considerations. A typical Carrousel extended aeration design summary for a 2-MGD train is shown in Figure 7.9.

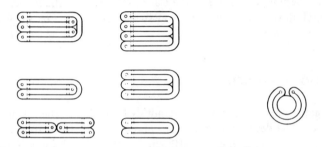

Figure 7.8. Carrousel tank configurations.

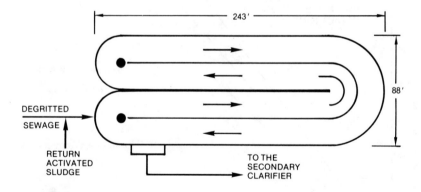

Figure 7.9. Typical Carrousel design.

INFLUENT CHARACTERISTICS	EFFLUENT CHARACTERISTICS	CARROUSEL CHARACTERISTICS	
FLOW = 2 MGD	BOD = 10 mg/l	BASIN VOLUME	1.6 MG
BOD = 200 mg/l	NH3N = 1 mg/l	CHANNEL WIDTH	22 FT
TKN = 25 mg/l		CHANNEL DEPTH	11 FT
INERTS = 0.25		NUMBER OF AERATORS	2
TEMP. = 10°C		AERATOR HORSEPOWER	50

Figure 7.10 shows the energy savings for a CLR using convective mixing as a function of mixing requirements and aerator oxygenation power. The other diagonal curves represent the conventional system mixing power requirements as a function of the inert solids level and SRT. As the wastewater influent inert solids fraction and SRT increase, the mixing requirements for a conventional surface aerator system increase, resulting in a greater energy savings for a Carrousel system alternative.

Many of the advantages purported for the Carrousel system are common to all CLRs. The unique feature of the Carrousel system is the harnessing of the horizontal flow component of a low-speed surface areator to produce convective mixing in a CLR.

JET AERATION CHANNEL

In the late 1960s LeCompt and Mandt [6] developed the jet aeration channel (JAC). The development was the culmination of efforts to treat large volumes of pulp and paper industry effluent. In 1968 the idea was conceived to use horizontally discharging ejectors to oxygenate and propel an aeration

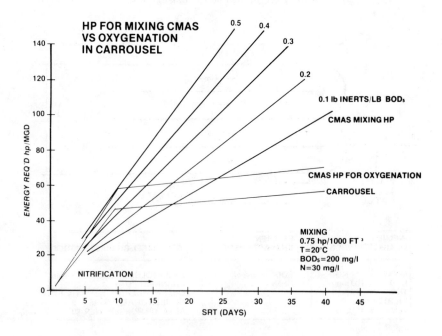

Figure 7.10. Mixing power requirements.

channel, overcoming the depth limitation of brush rotors [7]. In 1969 the concept was piloted on a coated-paper mill effluent [6] to evaluate biological performance. In 1970 a second pilot facility was used to determine propulsion and oxygenation efficiencies [6]. Later, Wilson and Friesen [8] made improvements to ejector aerators, developing a larger jet aerator; and Mandt [9] developed a multiple-unit manifold which found widespread use in JACs applied to both industrial and municipal wastewaters [10]. The JAC represents a significant American contribution to the art of CLR design [11,12].

In a jet aeration channel a mixture of air and mixed liquor emanating from horizontal jets is used to propel and aerate the fluid in the channel (Figure 7.11). Both pumps and compressors are employed to produce the plumes issuing from the jets, which are mounted near the bottom of the channel. The jets can be used for propulsion only (no air) to provide low DO or mixed anoxic conditions.

As compared with other looped reactor systems, the JAC has several advantages: depth and width of the channel are independent of each other, greater channel depths may be employed, the adverse cooling effect of surface aerators is avoided, and return sludge and raw sewage are admixed very efficiently with the mixed liquor via the jets.

In 1973 the first full-scale JAC was applied to whole black liquors from kraft pulp cooks and neutral sulfate cooks and the effluent from a business paper mill [12]. The plant was designed to handle 8500 lb BOD_5/day at a hydraulic rate of 3 MGD. Aeration volume was set at 2.2 MG, yielding a loading rate of 28.9 lb/100 ft^3/day. Design channel velocity was 1 ft/sec. Design clarifier rise rate was 530 gpd/ft^2. The channel was concentric about the final clarifier. No primary clarifier was used. The channel had an 85-ft i.d. and a 165-ft o.d. Channel width was 37 ft, 4 in. with a design water depth of 20.5 ft. Seven 8-jet manifolds (56 jets) located 3 ft off channel bottom provided oxygen and propulsion.

Example 7.1. Estimate the channel velocity of the above channel when each jet is fed 100 gpm of liquid and 30 scfm of air and the mixing nozzle diameter is 0.16 ft.

$$\text{momentum flux per jet} = \frac{Q\, \gamma_m\, V}{g} \tag{7.3}$$

$$\text{where } Q = \text{air water flow, ft}^3/\text{sec} = \frac{100}{60(7.48)} + \frac{30}{60} \times \frac{14.7}{22.3} = 0.553$$

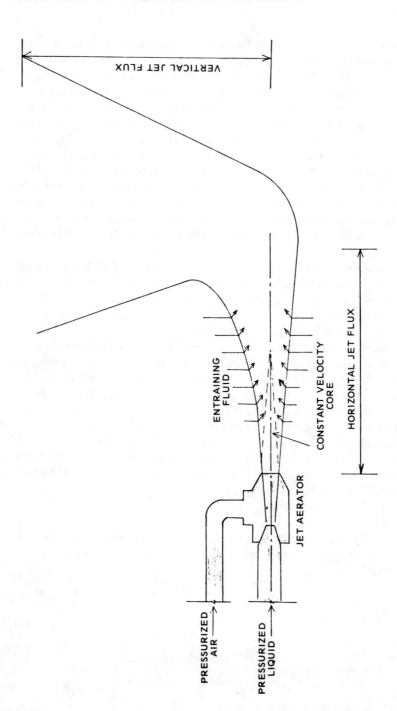

Figure 7.11. Jet aerator mixing pattern.

$$\gamma_m = \text{mixture density, air-water, lb/ft}^3 = \frac{0.223(62.4) + 0.3301(1.035)}{0.553} = 25.8$$

$$V = \text{velocity from mixing nozzle, ft/sec} = \frac{0.553}{\frac{\pi}{4}(0.16)^2} = 27.5$$

The momentum flux from the total numbered jets creates a small head, y:

$$y = \frac{\Sigma Q \ \gamma_m \ VJ}{gA \ \gamma_w} \tag{7.4}$$

where J = number of jets = 56
A = channel cross-sectional area, ft^2 = 37.33 x 20.5 = 821.3
γ_m = mixture density, lb/ft^3 = 25.8
γ_w = water density, lb/ft^3 = 62.4

$$y = \frac{56(0.553)(25.8)(27.5)}{(32.2)(821.3)(62.4)} = 0.0133 \text{ ft}$$

This head is available to overcome friction and to meet bend losses. The frictional loss, h_f, is estimated from Manning's equation:

$$h_f = \frac{n^2 L V^2}{(1.49)^2 \ R^{4/3}} \tag{7.5}$$

where n = Manning's number = 0.014 for concrete
L = channel length, mean = $\pi(85 + 37.33)$ = 384.3 ft
V_c = channel velocity
R = hydraulic radius = $\dfrac{\text{channel area}}{\text{wetted perimeter}}$ = $\dfrac{821.3}{(37.33) + 2(20.5)}$ = 10.58 ft

The bend losses, h_c, can be estimated from the data by Shukry published by Chow [13]:

$$h_c = fc \ x \ \frac{V_c^2}{2g} \tag{7.6}$$

where h_c = coefficient of curve resistance
fc = 1 for the full 360° turn of the channel

Equating the losses to the available head yields the velocity estimate:

$$y = \left[\frac{n^2 L}{(1.49)^2 R^{4/3}} + \frac{fc}{2g} \right] V_c^2 \qquad (7.7)$$

$$V_c = \left[\frac{0.0133}{\dfrac{(0.014)^2 \ 284.3}{(1.49)^2 \ (10.5)^{4/3}} + \dfrac{1}{64.4}} \right]^{1/2} = 0.9 \ \text{ft/sec} \qquad (7.8)$$

The above estimate makes no allowance for momentum transfer losses from aerator or propulsion devices. LeCompte [12] has defined losses associated with vertical pumping by jet bubble plumes. He has further defined a single head loss coefficient, C_L, which combines friction, curve, and plume losses. His suggested formula is:

$$C_L = \frac{y}{V^2} \ 2g \qquad (7.9)$$

He presented a plot of C_L vs air rate (Figure 7.12) for the channel described in the above example. Note that C_L is specific to the described channel. It is apparent that there is an optimum gas-to-liquid flow ratio for minimum C_L or maximum V_c beyond which bouyant plume back-pumping decreases velocity. (See Figure 7.13.) Note also that mixed liquor results can be as much as 20% less than clean water results.

The largest continuous looped reactor in the western hemisphere is a JAC plant treating pulp mill wastewater from Weyerhauser Company at Rothschild, Wisconsin. This plant treats up to 160,000 lb BOD_5/day. The largest municipal plant is in Linz, Austria, and treats up to 2 m^3/sec. The plant is also the world's deepest continuous loop reactor, with an operating depth of 7.5 m.

Oxygen transfer efficiencies of JACs are excellent. For JACs employing horizontal jets the transfer efficiency increases with increasing jet submergence. The energy efficiency tends to increase in JACs up to 10 ft of channel depth and remain more or less constant thereafter.

JACs employing inclined jets are particularly advantageous in that jet submergence can be reduced considerably while maintaining high transfer efficiency. The discharging jet plumes are used to carry gas bubbles downward in the channel for full oxygen dispersion. Blower pressure requirements are thus reduced and energy efficiency is increased. Figures 7.14 and 7.15 give oxygen transfer efficiency and energy efficiency for a typical JAC as a function of jet air-to-water volumetric flow ratio.

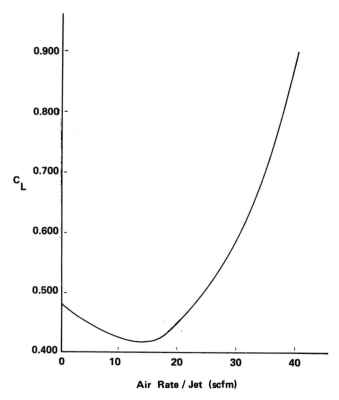

Figure 7.12. C_L vs air rate.

Example 7.2. Calculate the number of jet nozzles required to oxygenate a JAC with 150 lb/hr standard oxygen requirement, SOR, at a gas flow rate of 20 scfm per nozzle and an absorption efficiency of 30%.

$$SOR = Q_a \ E_o \ K \qquad\qquad (7.10)$$

where Q_a = total gas flow rate, scfm
  E_o = standard transfer efficiency, %/100
  K = conversion factor = 1.035 lb O_2/scfm

$$Q_a = \frac{150}{(0.3)(1.035)} = 485 \text{ scfm}$$

number of jets, j = 485 scfm ÷ 20 scfm/jet = 24 jets

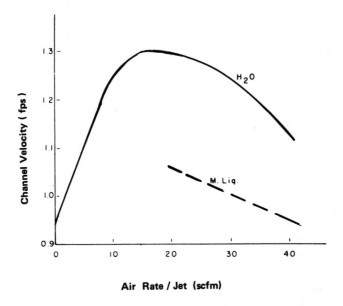

Figure 7.13. Channel velocity vs air rate.

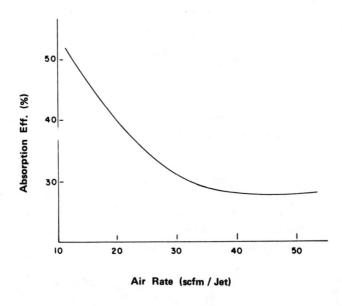

Figure 7.14. Absorption efficiency vs air rate.

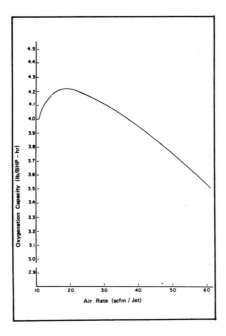

Figure 7.15. Energy efficiency vs air rate.

High-quality effluents are consistently reported for JACs [12,14]. High-quality effluents are attributable to full reactor volume utilization and good mixing as well as oxygenation provided by the continuous looped reactor/jet aerator combination.

ORBAL PROCESS

The orbal process, a multichannel oxidation ditch system, in which perforated aeration discs are the horizontal rotary devices used for oxygen transfer and mixing, was conceived in South Africa by Huisman and was researched and developed by the South African National Institute for Water Research. After lengthy test studies had been conducted and several small-scale plants had been in successful operation for several years, design technology was transferred to Envirex Inc. for promotion in the United States. Envirex initiated their marketing efforts in 1970. Presently, there are 70 orbal plants operating in the United States, ranging in size from 50,000 gpd to 9.6 MGD.

In the United States, several design changes were made, allowing the system to be economically attractive for large plants. The basin geometry was altered and channel depths were increased to a maximum of 12 ft. Additional design technology on nitrification and denitrification was obtained and incorporated into the designs where the extended aeration mode was used.

Orbal Basin Configuration

The typical orbal basin is oval-shaped and has two or more concentric aeration channels. The influent waste progresses in series from one channel to the next through submerged transfer ports. Each channel is a complete mix reactor having an endless flow circuit, allowing the influent flow to be quickly dispersed within the mixed liquor. The mixed liquor is aerated and effectively mixed by rotating aeration discs. Depending on the detention time and overall channel length, each individual flow particle will make numerous circuits before exiting to the next channel. The arrangement of complete mix reactors in series allows for a combination of complete mix/ plug flow benefits.

Aeration Discs

The rotating aeration discs used in the orbal are 4.5 ft in diameter and made of a noncorrosive plastic material that is split into two half sections for easy removal or attachment to the aerator shafting. The oxygenation capacity of the aeration disc varies with the immersion level and rotation speed. From the known oxygenation capacity of an individual aeration disc, it is possible to obtain the oxygenation intensity in each particular shaft section. Shafts, spanning one or more channels, drive sets of discs in each channel from a common gear drive.

Process Modes

The multiple-channel arrangement allows for different process modes to be used. A conventional activated sludge process mode can be used, with the inner channels used for aeration and the outside channel used for aerobic digestion. Contact stabilization and step aeration can also be used by directing the influent and return sludge flows to the appropriate channels. To assure consistent high-degree treatment and reduce sludge volume, the extended aeration mode is also used when nitrification is a requirement.

Nitrification-Denitrification

Applegate [15] showed that it was possible to obtain complete nitrification in a multichannel orbal while still achieving a high degree of denitrification. The performance level measured in the plant of his study indicated effluent ammonia levels near zero and effluent nitrate levels between 1 and 4 mg/L. Nitrification performance in an orbal is optimized because of the elimination of ammonia short-circuiting, while denitrification performance is consistently good because DO levels are continually kept near zero in the first channel reactor.

In an orbal system extended aeration process, there is typically a DO stratification across the series of channels. In a three-channel system, the DO level of the first channel is typically near zero, the second channel between 0.5 and 1.5 mg/L and the third channel between 1.5 to 3.0 mg/L. Denitrification and nitrification *both* take place in the first channel under zero DO conditions. The degree of nitrification (and BOD removal) that takes place in the first channel depends on the amount of oxygen delivered. With the oxygen uptake rates in the first channel being typically high, up to 90% of the process oxygen requirements can be delivered in this reactor, while still maintaining the DO level at zero. Lower oxygen uptake rates keep the DO level in the later channels high despite lower oxygen delivery rates in these reactors. Depending on the ratio of BOD to ammonia and the percent process work accomplished in the first channel, the oxygen delivery savings allowable through denitrification and the zero DO level of the first channel is approximately 35%.

Design Parameters for Orbal

The extended aeration orbal is designed typically for a sludge age of 20 to 30 days with a MLSS concentration between 4000 and 6000 mg/L and a BOD loading rate of 12 to 20 lb/1000 ft^3. After the aeration volume has been determined, a channel depth between 6 and 12 ft is selected, and a channel volume distribution is made in which the first channel has 50 to 70% of the total volume. For simplicity of aeration equipment design, the channels of one basin have similar widths. Channel depths do not exceed the channel width. Aerator shaft section lengths do not exceed 20 ft, but multiple shaft sections can be used in single channels.

The orbal basin configuration is typically oval in shape. To reduce overall wall lengths and optimize channel widths, approximatey 80 to 90% of the total tank volume is kept in the curved portion of the channels and the straightaway sections are kept as short as possible.

Oxygen requirements are based upon the varying process work performance from channel to channel and the design DO stratification. In a three-channel system having a selected volume split of 50/33/17, and a DO stratification of 0/1/2, the oxygen delivery requirements are calculated as follows:

Given: X   =   actual oxygen needs (with alpha, beta, and temperature corrections

volume split   =   50/33/17

DO strati-
     fication   =   0/1/2 (correction factors of 1/0.89/0.78)

process split   =   65/25/10 (dependent on volume split)

Oxygen Delivery

First Channel	0.65 x/1 = 0.65x
Second Channel	0.25 x/0.89 = 0.28x
Third Channel	0.10 x/0.78 = 0.13x

The aeration system has performance characteristics that are dependent on rotation, immersion and direction. At 55 rpm and 21-in. immersion, each aeration disk will deliver 1.85 lb O_2/hr at 3.2 lb O_2/hp-hr efficiency in one direction and 2.4 lb O_2 at 3.0 lb O_2/hp-hr efficiency in the alternate rotation direction.

In the allotment of aeration disks to the shaft sections, a spacing of at least 9 in. is required. After determining channel widths and oxygen requirements per channel, the number of discs per shaft section and number of assemblies per basin can be calculated. Size of motors can be determined from number of discs per assembly and disc efficiency rating. Motor size selection also should include a 10% reserve figure for starting purposes.

Mixing requirements may govern under certain conditions. With the disc aerators in a multiple-channel basin, approximately 1 hp of discs is required to mix 50,000 gal of mixed liquor and keep the solids in suspension. A typical orbal plant is shown in Figure 7.16.

PASVEER-TYPE PLANTS

Continuous loop reactor plants resembling the original Pasveer ditch use single loops employing horizontal rotor-aerators. The oxygenation and mixing capacity of the rotor is adjusted by manipulation of adjustable weirs installed at the ditch effluent, which control the liquid level and rotor immersion. The initial ditch plants were elongated ovals with sloping side walls and center island. In addition to the oval configuration, C-shaped and U-shaped configurations are employed. Liquid operating depths generally range from 3 to 6 ft. The development of larger-diameter (42 in.) rotors (by Lakeside) has allowed an increase in operating depth, peaking at 10 to 14 ft. At liquid operating depths greater than 7 ft, a horizontal baffle downstream of the rotor deflecting flow toward the channel bottom is required in order to get full depth mixing.

In the United States, the volume of the ditch is generally sized based on a loading of 10 to 35 lb BOD_5/1000 ft^3 of reactor volume. A minimum hydraulic retention time of 18 to 24 hr is commonly set regardless of organic loading.

Figure 7.16. Typical orbal plant (courtesy of Envirex, Inc.).

Normally the Pasveer type CLR has sloping side walls. A compacted earthen ditch is preferably lined with 4 to 6 in. of poured concrete or shotcrete. Asphalt, wood, preformed materials and clay have also been used. The rotor moves the liquid in a horizontal plane in such a fashion that it is preferable to have a straight section at least 40 ft long directly downstream from the discharge of all rotors. Where center islands are used, they should be wide enough to provide a smooth flow around the bend. The width of the island should vary depending on the width of the ditch at the liquid level. Normally, a 12-ft center island should be used for ditch widths up to 13 ft, a 10-ft island for widths from 13 to 24 ft, and larger islands where the ditch width is 25 ft or greater.

Vertical side walls are also employed with a center island or center dividing wall. Consideration should be given to flow guide baffles or fluting at bends to reduce bend losses and minimize deposition of solids in the downstream lee of dividing walls coming out of bends. Raw waste and return sludge are preferably added just upstream of the rotor in order to thoroughly mix and distribute these flows across the cross section at the rotor and prevent short-circuiting.

Rotors

Rotors are mechanical surface aerators which rotate in a plane horizontal to the liquid surface and perpendicular to the channel. Present day rotors or "brushes" are second generation Kessener brushes. Kessener, a Dutch engineer, developed the first aeration rotor in 1926 for use in conventional activated sludge plants. Pasveer adapted the Kessener brush 25 years later for his oxidation ditch. Each rotor unit consists of a motor; a gear, belt or chain drive; and a horizontal pipe to which steel blades or tines are attached in a radial fashion. (See Figures 7.17 and 7.18.)

Rotor oxygenation rate per unit of length is a function of blade submergence and rotational speed (see Figure 7.19). Transfer rates of 3.5 lb per linear foot of rotor (5.2 kg/m) are common. A comparison of rotor devices by various manufacturers as summarized by the U.S. EPA [16] is reproduced in Table 7.3. Oxygen transfer efficiency is comparable to vertical turbine aerators and typically varies from 2.5 to 3.0 lb/hp-hr (1.5 to 1.8 kg/kWh). The length of rotors used for a given project is determined by the greater of the maximum length computed to maintain the desired ditch velocity or to transfer the peak oxygenation rate required. Ditch velocity is based on the propulsion capability of the rotor and the frictional and inertial (bend) losses of the ditch. Most Pasveer-type CLRs are designed for a minimum, mean horizontal velocity of 1 ft/sec (0.3 m/sec). One should recognize that it is not the mean horizontal velocity that is directly responsible for solids

Figure 7.17. Rotor aeration (courtesy of Lakeside Equipment Corp.).

Figure 7.18. Rotor aeration in concrete channel.

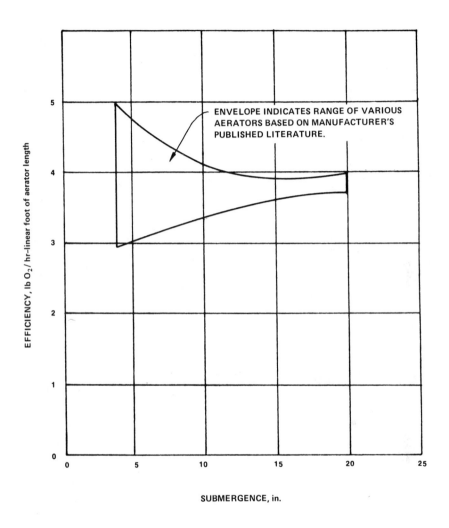

Figure 7.19. Rotor oxygenation rate.

suspension, but the secondary turbulence resulting from flow at the required velocity within the CLR.

Rotors are typically 50% efficient in transferring momentum from motor torque to ditch velocity. Thus, once the energy required for a particular CLR geometry and velocity is determined, the required motor power draw can be approximated as twice the hydraulic energy requirement to propel the CLR contents. Experience has shown [17] that a well configured Pasveer CLR requires 60 to 70 hp/mg (0.012 to 0.014 kW/m^3) to provide proper velocity

Table 7.3. Comparative Aerator Characteristics

Manufacturer	Type or Model	Type Blades	Diameter, in.	Shaft	Typical Submergence Range, in.	Typical Speed Range, rpm	Material of Construction	Max. Length per Shaft, ft
Lakeside	Cage & Mini	Horizontal-toothed blades	27½	6-in.-o.d. torque tube	2–10	60–90	steel	16
Lakeside	Magna	3-in.-wide brush, 14-in.-long max.	42	14-in.-o.d. torque tube	4–14	50–72	steel	30
Passavant	Manmoth Series 5300	3-in.-wide brush, 9.5-in. long	27	8-in.-o.d. torque tube	3–9.5	90	steel	12
Passavant	Manmoth Series 5300	3-in.-wide brush, 12-in. long	39-3/8	14-in.-o.d. torque tube	4–12	70	steel	30
Walker	Reelaer Class 6227	3-in.-wide brush, 11-in. long	38	16-in.-o.d. torque tube	7–10		steel torque tube, galvanized blades and hardware	12 min. 25 max.
Envirex	Disc	1/2-in.-wide plastic disc	52	5-in.-o.d. torque tube	11–21	56–58	steel torque tube	
Cherne	OTA Aerator	Perforated fiberglass blades	30			≤110	fiberglass	7 (one size)

and mixing. Lakeside [17] recommends a maximum CLR volume per lineal foot of rotor as given in Table 7.4.

In 1975, Lakeside had 557 CLR installations in the U.S. and Canada, Passavant 56, and Cherne 1 [16]. Figures 7.20 through 7.23 show typical Pasveer CLRs.

BARRIER DITCHES

Barrier ditches are not true CLRs in that conservation of mixing energy and flow momentum is not practiced. Rather, a barrier spans the cross section of a channel and flow is pumped from the upstream side of the barrier to the downstream side. A down-pumping, sparged turbine and draft tube is employed in Lightning's draft tube channel, Figure 7.24. A surface aerator and up-pumping draft tube is used in EPI's Channel Air system.

Oxygenation efficiencies of barrier ditches are comparable to the aeration devices employed, i.e., surface mechanical aerators and sparged turbines. The sparged turbines used employ an axial flow impeller (down-pumping) and an air compressor. Air is introduced through a sparger ring below the turbine impeller. The pumped flow carries the air down the draft tube for discharge on the downstream side of the barrier. The barrier prevents backflow in the channel and allows for a free surface rise in the channel to create a head differential between the upstream and downstream sides of the barrier. This head differential creates flow in the channel. The proper ratio of turbine pumping to air introduction is required so as not to overgas the turbine.

The propulsion requirements for the channel are easily determined once the mechanical efficiencies of the turbine, the head loss through the turbine/ draft tube system, and the channel losses or free surface rise are known or calculated. The total head loss, h_T, through the system is given by Equation 7.11.

Table 7.4. Rotor Mixing Capabilities

Rotor Diameter, in.	Maximum CLR Volume, gal/ft of rotor
27.5	13,000 plants < 600 P.E.[a]
27.5	16,000 plants > 600 P.E.
42	21,000

[a]P.E. = population equivalent.

Figure 7.20. Concentric Pasveer CLRs.

Figure 7.21. Multirotor CLRs.

Figure 7.22. Folded Pasveer-type CLR.

Figure 7.23. U-shaped Pasveer-type CLR.

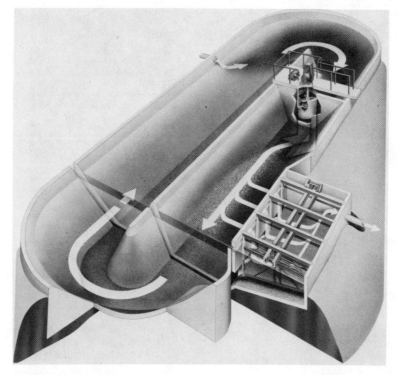

Figure 7.24. Draft tube channel barrier ditch (courtesy of Mixing Equipment Co.).

$$h_T = h_{CLR} + h_{TURB} \tag{7.11}$$

where h_{CLR} = losses in channel (due to friction and bends)
h_{TURB} = turbine and draft tube losses

The energy, E_p, required for propulsion can be calculated as follows:

$$E_p = \frac{h_T \, Q_{CLR}}{\eta_p \eta_d \eta_m \, 65.95} \tag{7.12}$$

where h_T = total head losses, ft
Q_{CLR} = total flow in CLR, ft^3/sec
η_p = pumping efficiency
η_d = drive efficiency
η_m = motor efficiency

Example 7.3. Determine the required propulsion energy for a barrier ditch with a cross-sectional area of 200 ft² and a required mean channel velocity of 1 ft/sec when the pumping efficiency at the design gassing rate is 70%, the drive efficiency is 94%, and the motor efficiency is 90%. Head loss around the channel is 0.3 ft and the head loss through the draft tube is 5 ft, including draft tube entrance and exit losses.

$$Q_{CLR} = V_{CLR} \times A_{CLR} = 1 \text{ ft/sec} \times 200 \text{ ft}^2 = 200 \text{ ft}^3/\text{sec} \qquad (7.13)$$

$$E_p = \frac{5.3 \ (200)}{(0.7) \ (0.94) \ (0.9) \ 65.95} = 27.1 \text{ hp} \qquad (7.14)$$

COMBINED SYSTEMS

A number of CLR-type systems have recently emerged employing either rotating diffusers or fixed diffusers with separate propulsion devices to create circulation. Scriebner offers a system using rotating diffusers which create channel velocity as they rotate. The system is limited to circular channels that can accommodate the rotating arms. Fluidyne has offered a CLR system using fixed fine bubble diffusers and a submerged horizontal propeller or jets as well as a system using an airlift and surface jets for oxygenation and propulsion.

Unfortunately, these systems have not been in existence long enough to generate meaningful operation and performance data and history. With the continued and growing interest in CLR technology, new concepts and innovations in CLR methodology, process and equipment are certain to evolve with time. Each should be evaluated in the light of the information presented in this text. Each should be evaluted after consultation with the process or equipment innovator.

REFERENCES

1.   Kout, A. C. J., and J. Keper, "Carrousel, a New Type of Aeration System with Low Organic Load," *Water Res.,* 6:401 (1972).
2.   Jacobs, A., "A New Loop in Aeration Tank Design," paper presented at the New York Water Pollution Control Association Meeting (June, 1964).
3.   Kalbskopf, J. H., "Requirements of Aeration and Circulation Systems Within the Scope of the Development of the Activated Sludge Process and Environmental Protection."

4. Downing, et al., "Nitrification in the Activated Sludge Process," *J. Proc. Inst. Sew. Purif.*, 63(2) (1964).
5. Stensel, H. D., and G. L. Shell, "Two Methods of Biological Treatment Design," *J. Water Poll. Control Fed.*, 46(2):271 (Feb., 1974).
6. LeCompte, A. R., "Ejector Aerated Oxidation Ditch for Waste Treatment by Channel Aeration Propelled and Oxygenated with Ejectors," paper presented at the 44th Annual Conference of WPCF (Oct., 1971).
7. LeCompte, A. R., "Ejector Aerated Oxidation Dtich for Waste Treatment," United States Patent 3,846,292 (1974).
8. Wilson, G. E., and M. E. Friesen, "Method of and Apparatus for Jet Aeration," United States Patent 4,112,025 (1978).
9. Mandt, M. G., "Multiple Jet Aeration Module," United States Patent 3,897,000 (1975).
10. Mandt, M. G., "Jet Aeration Channel System," United States Patent 4,199,452 (1980).
11. Huang, J. Y. C., and M. G. Mandt, "Jet Aeration Theory and Application," 28th Purdue Industrial Waste Conference, Lafayette, IN (1973).
12. LeCompte, A. R., "A Pasveer Ditch American Jet Style," 29th Purdue Industrial Waste Conference, Lafayette, IN (1974).
13. Chow, V. T., *Open Channel Hydraulics* (McGraw-Hill Book Company, 1959).
14. Wiskow, R. W., "Oxyditch Concept Proven in Pioneer Municipal Plant," *Public Works* (April, 1979).
15. Applegate, C. S., B. Wilder and J. R. DeShaw, "Total Nitrogen Removal in a Multi-Channel Oxidation System," *J. Water Poll. Control Fed.*, 52(3):568-577 (1980).
16. Ettlich, W. F., "A Comparison of Oxidation Ditch Plants to Competing Processes for Secondary Advanced Treatment of Municipal Wastes," U.S.EPA publication, 600/2-78-051 (March, 1978).
17. Berk, W. L., "Why Settle for Only Secondary Treatment," Lakeside Equipment Corp. Publication, RAD-390.

CHAPTER 8

OPERATION AND MAINTENANCE

The major advantages of oxidation ditch plants are their ease of operation and their reliability [1,2]. The ease of operation and reliability of oxidation ditch plants have been attested to by EPA: "Observations based on this study indicate that, as a group, oxidation ditch plants can be operated by average personnel to produce above average performance results [1]," and by experiences in England: "This carrousel plant is easy to operate with few mechanical or electrical problems [2]." It is apparent that the large biomass contained in an oxidation ditch, along with the long hydraulic and solids retention times used, contributes greatly to the stability of oxidation ditch plants. These factors alone, however, are not sufficient to explain the operational success of oxidation ditch plants. Extended aeration plants have the same characteristics of large biomass and long retention times but do not reliably produce high-quality effluents [1,3]. It may be conjectured that the unique combination of short-term plug flow (one loop) with overall complete mix process configuration leads to the overall reliability of oxidation ditch treatment processes.

OPERATIONAL REQUIREMENTS

While oxidation ditch processes have their unique operational characteristics, which will be discussed in this chapter, it must be recognized that they are a modification of the activated sludge process. As such, the overall operational considerations for activated sludge processes apply. It is not possible to discuss here all the operational requirements of activated sludge processes. For this information, the reader is referred to the excellent WPCF publication

129

on operation [4]. This chapter will cover only those operational characteristics applicable to oxidation ditches.

Control of Oxidation Ditches

The overall control of an oxidation ditch plant depends on biomass control and control over dissolved oxygen levels. Biomass control may be obtained in a number of ways. To maintain stable operation it is necessary that a constant biodegradable organic load be maintained at all times. The most common ways of achieving this end is through control of food-to-microorganism ratio (F/M) or solids retention time (θ_c). Both parameters may be adjusted as required to meet changes in waste characteristics or temperature.

The use of the F/M ratio requires knowledge of the organic load present as well as the biomass present. Real-time data regarding biodegradable organic loadings are not available, although this information can be approximated through the use of chemical oxygen demand (COD) or total organic carbon (TOC) measurements combined with BOD/COD or BOD/TOC correlations. The biomass is traditionally measured by assuming volatile suspended solids represent biomass. In oxidation ditch plants, primary clarifiers are rarely used. Therefore, a significant but variable fraction of the MLVSS measured may be nonviable organic matter. These two factors make control based on F/M ratio difficult.

Control through solids retention time requires the wasting of a constant percentage of the biomass each day. For example, to obtain a solids retention time of 20 days it is necessary to waste 5% of the biomass daily. Since the viable biomass will be wasted as a fixed fraction of the total solids wasted, control may be exerted simply by wasting the desired percentage of total ditch suspended solids. Control through use of solids retention time does not require the availability of real time organic loading data. For these reasons, control of oxidation ditches through use of solids retention time is recommended.

Control of waste activated sludge (WAS) requires facilities to measure WAS flow rate and concentration. Wasting of sludge directly from the oxidation ditch offers advantages in control and in further handling of the waste sludge [5,6].

Return activated sludge must be controlled to maintain the desired solids content in the oxidation ditch. Continuous withdrawal of sludge from the secondary clarifiers is recommended to prevent anoxic conditions in the sludge and the potential for rising sludge due to denitrification. The operator must be provided with information to determine solids content and flow rate of the return activated sludge.

Oxygen

The other major control parameter for oxidation ditch plants is the dissolved oxygen concentration in the reactor. Almost all oxidation ditch plants are designed to allow some control over oxygen input. This may be accomplished with variable submergence or speed on mechanical aerators and blower turndown when jet aerators are used. In plants designed for BOD removal and/or nitrification, control is utilized to maintain a dissolved oxygen level greater than or equal to 2 mg/L. Considerable cost savings through reduced aeration power requirements may be obtained through control to prevent unnecessarily high dissolved oxygen levels.

When denitrification is desired, close control over oxygen input is required in order to ensure the maintenance of anoxic zones. EPA [1] has considered the use of manual dissolved oxygen measurements to be sufficient to maintain control. The English experience [7] has been that location of a dissolved oxygen probe at the desired start of each anoxic zone has been more successful. Although the use of DO probes requires frequent cleaning and calibration of the probe, the use of probes is recommended for those systems used for denitrification.

Instrumentation

Except for dissolved oxygen control, the instrumentation required for the proper control of oxidation ditch plants is similar to the instrumentation required for activated sludge systems [1]. These requirements, modified from the EPA study [1] plus DO control are:

- raw sewage or effluent flow measurement, recording and totalizing
- return sludge measurement and recording
- waste sludge flow measurement and recording
- chlorine feed pacing from flow
- DO measurement, recording and aerator control
- normal laboratory instrumentation

OPERATION AND MAINTENANCE REQUIREMENTS

The operation and maintenance requirements of oxidation ditch plants are similar to activated sludge plants. Preventive maintenance, chlorine requirements, and disposal of screenings and grit are similar to other plants and will not be discussed here. Those items such as manpower, electrical and training requirements which are specific to oxidation ditches are discussed.

Operational and maintenance labor requirements as well as electrical power requirements for oxidation ditches are presented in Table 8.1. As may be seen from the table, little manpower is required to achieve denitrification and significant electrical power savings can be realized.

Shanks and Connell [8] have reviewed the role of operator training in the successful operation of oxidation ditch plants. They point out that operation contributes more to the success or failure of a plant than does design. Good operation depends on: the availability of a well written and complete operation and maintenance (O&M) manual, cooperation of the design engineer with the operation staff, and hands-on and seminar training. This need not be overly expensive. At one oxidation ditch facility, Shanks and Connell [8] found that the training costs were less than 6% of the first year's cost.

PERFORMANCE OF OXIDATION DITCH PLANTS AND COMPETING PROCESSES

As indicated earlier in this chapter, oxidation ditch plants are capable of consistently producing excellent-quality effluents with a minimum of operation. In this chapter, both average effluent quality and reliability of performance will be examined.

Table 8.1. O&M Requirements for Oxidation Ditch Plants [1]

Type and Flow, MGD	O&M Labor, Man-days/yr	Electric Power, 1000 kWh/yr
BOD Removal and Nitrification		
0.05	226	46
0.10	257	72
1.00	388	280
5.00	1911	2000
10.00	3818	3700
Denitrification		
0.05	310	37
0.10	340	58
0.50	490	224
1.00	630	400
5.00	2030	1600
10.00	3930	2960

Oxidation Ditches

The overall performance of oxidation ditch plants is summarized in Table 8.2. These data, taken from 29 plants, indicate that in terms of organic and suspended solids removal, oxidation ditch plants are capable of producing an average effluent of less than 15 mg/L BOD_5 and TSS. Data obtained from the same study indicate the high reliability of oxidation ditch plants. Table 8.3 presents reliability data obtained from the EPA study [1]. In Tables 8.2 and 8.3 it should be noted that the results presented are primarily from plants not designed for nitrogen removal. On the average, oxidation ditch plants are also capable of producing effluents containing less than 2 mg/L of ammonia.

Similar operational results have been obtained outside the United States. Table 8.4 presents the results from the Cirencester plant in England. Results from the Cirencester plant are excellent despite the large variation in influent

Table 8.2. Oxidation Ditch Plant Performance [1]

	Effluent, mg/L			Removal, %		
	Winter	Summer	Total	Winter	Summer	Total
BOD_5						
High Plant	55	34	41	87	86	87
Average	15.2	1.2	12.3	92	94	93
Low Plant	1.9	1.0	1.5	99	99	99
Suspended Solids						
High Plant	26.6	19.4	22.4	81	82	82
Average	13.6	9.3	10.5	93	94	94
Low Plant	3.1	1.9	2.4	98	98	98
Total Nitrogen (TKN)						
High Plant	10	6.6	8.1	55	61	57
Low Plant			3.0			72
Ammonia Nitrogen						
High Plant	7.2	1.03	3.3	81	97	91
Average	3.8	0.6	1.8	90	98	99
Low Plant	0.4	0.1	0.3	98	99	99
Nitrate Nitrogen						
High Plant	29	33	31			
Average	12.5	15.1	14.1			
Low Plant	5.6	4.5	5.0			

Table 8.3. Reliability of Oxidation Ditch Plants [1]

	Percent of Time That Effluent Concentration is Less Than:								
	10 mg/L			20 mg/L			30 mg/L		
	TSS	BOD$_5$	Total Nitrogen	TSS	BOD$_5$	Total Nitrogen	TSS	BOD$_5$	Total Nitrogen
Best Plant	99	99		99	99		99	99	
Average of All Plants	65	65	40	85	90	69	94	96	88
Worst Plant	25	25		55	55		80	72	

quality. At no time was 30 mg/L BOD$_5$ exceeded in the effluent and 95% of the time the effluent contained less than 15 mg/L TSS and 10 mg/L BOD$_5$. Oxidation ditch plants are capable of meeting the current definition of secondary treatment of 30 mg/L BOD$_5$ and TSS 94% of the time. The worst plant studied was able to meet the 30 mg/L BOD$_5$ and TSS standard over 70% of the time. As impressive as these results are, comparison with other biological treatment processes is necessary to fully appreciate the performance of oxidation ditch plants.

Competing Processes

Results for the commonly used biological treatment processes are presented in Table 8.5 and the reliability data for these processes are presented in Table 8.6. Both Tables 8.5 and 8.6 present results for the average of these types of plants. None of those processes have been shown to be able to produce an average effluent equal to that produced by oxidation ditches. Neither is the reliability of competing processes as good as the reliability shown by oxidation ditches. Comparing average plants, oxidation ditches can produce a 30 mg/L BOD$_5$ and TSS effluent 94% of the time while the best competing process, activated sludge, can produce this quality of effluent only 85% of the time, and no competing process can meet this effluent quality as reliably as the worst oxidation ditch plant studied.

Activated sludge reliability for large-scale and package plants is presented graphically in Figures 8.1 through 8.3. Oxidation ditch plant reliability is also shown on these figures for comparison. In Figures 8.2 and 8.3 temperature effects on package plant performance may be clearly seen. The Cincinnati area plants operate at both summer and winter temperatures; the Dade County plants operate only in warm weather.

Table 8.4. Operational Results at Cirencester Carrousel Plant, Daily Means [2]

	Raw Sewage			Plant Effluent				
	BOD$_5$, mg/L	TSS, mg/L	NH$_3$-N, mg/L	BOD$_5$, mg/L	TSS, mg/L	NH$_3$-N, mg/L	NO$_2$-N, mg/L	NO$_3$-N, mg/L
Annual Average	101	169	16.4	1.9	5.8	1.4	0.2	4.7
Minimum	10	31	4.0	trace	trace	trace	trace	trace
Maximum	400	686	39.3	30	224	6.5	1.3	13A
Not Exceeded 95% of the Time	216	299	31.2	7.8	12.3	4.4		

Table 8.5. Competing Biological Processes Performance [1]

	Effluent, mg/L		Removal, %	
	TSS	BOD$_5$	TSS	BOD$_5$
Activated Sludge (1.0 MGD)	31	26	81	84
Activated Sludge (Package Plants)	28	18		
Trickling Filters	26	42	82	79
Rotating Biological Contactor	23	25	79	78

Table 8.6. Competing Biological Processes Reliability [1]

	Percent of Time Effluent Concentration Less Than:					
	10 mg/L		20 mg/L		30 mg/L	
	TSS	BOD$_5$	TSS	BOD$_5$	TSS	BOD$_5$
Activated Sludge (1.0 MGD)	40	25	75	70	90	85
Activated Sludge (Package Plants)	15	39	35	65	50	80
Trickling Filters		2		3		15
Rotating Biological Contactor	22	30	45	60	70	90

The comparison with other processes demonstrates the ability of oxidation ditch plants to consistently produce a higher-quality effluent than other commonly used processes. The excellent performance of oxidation ditch plants was obtained with only average operational input.

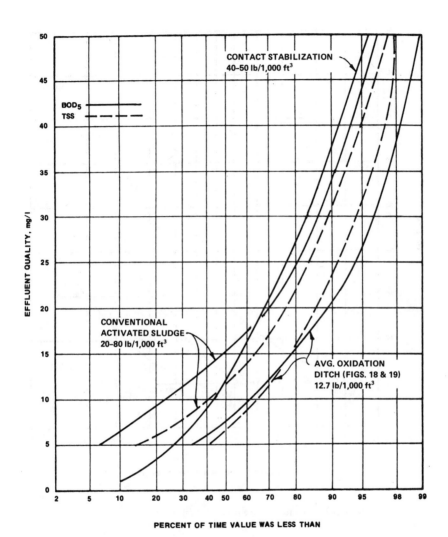

Figure 8.1. Activated sludge effluent quality [1].

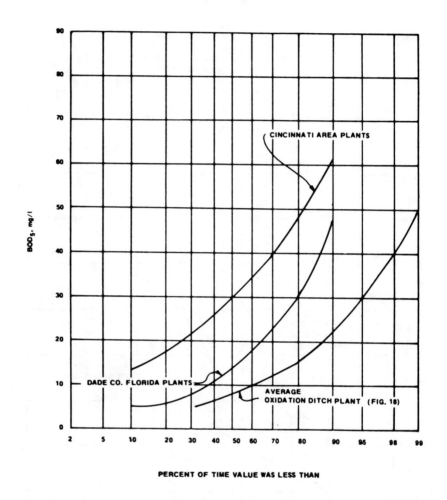

Figure 8.2. Activated sludge package plant reliability, BOD$_5$ [1].

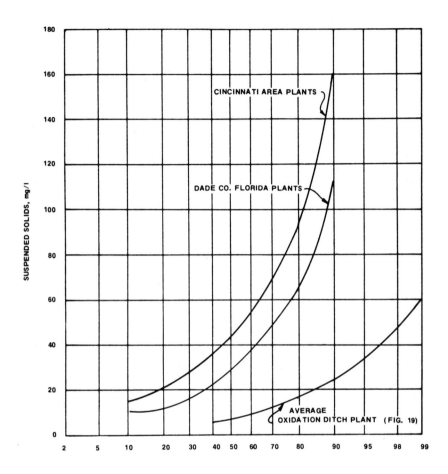

Figure 8.3. Activated sludge package plant reliability, SS [1].

REFERENCES

1. "A Comparison of Oxidation Ditch Plants to Competing Processes for Secondary and Advanced Treatment of Municipal Wastes," U.S. EPA, EPA-600/2-78-051 (1978).
2. Carmichael, W. F., and D. M. W. Johnstone, "Initial Operating Experiences at Cirencester Carrousel Plant," Presented at Joint Meeting Inst. Wat. Pollution Control and Inst. Pub. Health Engrs., Cirencester, England (1980).

3. Guo, P. H. M., D. Thirumurth and B. E. Jank, "Evaluation of Extended Aeration Activated Sludge Package Plants," *J. Water Poll. Control Fed.*, 53(1):33-42 (1981).
4. "Operation of Wastewater Treatment Plants," Water Pollution Control Federation, MOP 11, Washington, DC (1976).
5. Metcalf and Eddy, Inc., *Wastewater Engineering,* 2nd edition (New York: McGraw-Hill Book Company, 1979).
6. "Sludge Thickening," Water Pollution Control Federation, MOP FD-1, Washington, DC (1980).
7. Hanbury, M. J., A. J. Rachwal, D. W. M. Johnstone, D. Critchard and G. C. Cox, "Evaluation of the Carrousel System and Its Potential for Denitrification," presented at Inst. Wat. Engrs. and Sci., Ash Vale, England (1978).
8. Shanks, R. L., and J. E. Connell, "Operator Training is Key to Oxidation Ditch Start-Up and Operation," *J. Water Poll. Control Fed.*, 53(4):444-450 (1981).

CHAPTER 9

DESIGN OF CONTINUOUS LOOP REACTORS

In this chapter we will discuss the design of treatment facilities utilizing oxidation ditches for biological treatment. It is, of course, not possible to discuss the entire scope of design involved for a complete treatment plant in one chapter. We will concentrate on those aspects of design which are directly affected by the use of oxidation ditches. These include: the oxidation ditch, oxygen transfer equipment, clarifiers and sludge disposal alternatives. Other aspects of the treatment plant will either be mentioned briefly or not at all. For an excellent starting point on other aspects of design the reader is referred to MOP 8, published by the Water Pollution Control Federation [1]. It has been assumed that wastewater volumes and characteristics have been determined. This process will not be reviewed here.

Included in this chapter is a complete design example. It is presented since the authors feel that verbal descriptions of the design processes are always inadequate. It is not intended that the design example be taken as the only way to design the plant, but rather as an example of a possible approach to design.

PRELIMINARY AND PRIMARY
TREATMENT

Oxidation ditch plants require the same preliminary treatment facilities as other types of treatment systems. Pumping of influent is often required. For design of influent pumping systems the reader is referred to several excellent references in the literature [1-3]. If the collection system is long, resulting in long detention times, the wastewater may arrive at the plant in a

141

septic condition. Prechlorination and/or preaeration may be used in these cases for odor control and to prevent the formation of corrosive atmospheres.

Coarse screening of influent wastewaters is included in most treatment facilities, primarily to remove large objects which may damage or clog plant equipment or piping. In medium to large plants (> 2 MGD), the bar screens are almost always mechanically cleaned, whereas in smaller plants they may be mechanically or manually cleaned [1].

Grit removal may or may not be included in oxidation ditch plants. In general, grit removal is carried out to protect equipment and piping from abrasion and clogging, as well as to prevent the filling of anaerobic digesters with grit. In oxidation ditch plants, anaerobic digesters are rarely used. Thus, grit removal will not be needed to prevent digester filling. Whether or not to use grit removal is essentially an economic decision. The cost of grit removal facilities and their operation must be evaluated against the cost of grit-resistant pumps and increased operation and maintenance costs due to abrasion and clogging.

Primary clarification is rarely used with oxidation ditches in the treatment of municipal wastes. Oxidation ditches, particularly when operated in the extended aeration range, are capable of economically treating municipal wastes with no primary treatment. Since the extended aeration process produces a stabilized sludge, the omission of primary clarifiers (sludge) is very attractive.

When an oxidation ditch is to be designed in the conventional activated sludge range, or when the wastewater has a very high solids concentration, primary clarifiers may well be used. In these systems, grit removal and sludge handling alternatives must be carefully considered.

REACTOR DESIGN

The aspects of reactor design discussed in this chapter are general and should apply to all types of continuous loop reactors. For details particular to any of the proprietary processes discussed in Chapter 7, the manufacturer supplying the process should be consulted.

Kinetic Design

The details of kinetic design of the reactor have been presented and discussed in Chapter 2 for carbonaceous BOD removal and in Chapter 4 for nitrification and denitrification. We will summarize the steps involved in the kinetic design and sizing of the reactor.

For those systems designed for carbonaceous BOD removal, only the following procedure may be followed:

1. Determine influent wastewater characteristics and effluent requirements.
2. Ensure that pH and nutrient levels in the influent are suitable for biological wastewater treatment.
3. Determine required effluent soluble BOD_5 by the method presented in Chapter 2 (see Example 2.3).
4. Select the design solids retention time. Note that if stabilized sludges are desired, θ_c greater than 20 to 30 days will be required. If not, θ_c values of 5-8 days at 20°C are common for BOD removal. Recommended ranges of θ_c can be found in the literature [1, 4-9].
5. Determine from pilot studies or literature values for the growth yield constant, Y, and the endogenous decay coefficient, k_d. For typical values see Tables 2.3 and 2.4.
6. From Equation 2.61, calculate a value for ditch volume times MLVSS concentration.
7. Select a value for MLVSS concentration. Common values for MLVSS concentration in oxidation ditches range from 1500 to 4000 mg/L. Remember that at high MLVSS concentrations, the clarifier design will be controlled by the thickening characteristics of the clarifiers.
8. Calculate the volume of the reactor and the hydraulic retention time.
9. Calculate the return activated sludge rate using a simple mass balance around the ditch, as shown later in this chapter.
10. Determine the oxygen requirements as discussed later. After determination of AOR, aerators may be sized and selected as described in Chapter 5.
11. Clarifiers may be sized and waste sludge calculated as discussed later in this chapter.

When nitrification and/or denitrification are required, a slightly different kinetic design approach is followed. The following procedure is recommended:

1. Determine influent wastewater characteristics and effluent requirements.
2. Ensure that influent pH and nutrient levels are suitable for biological treatment.
3. Estimate the TKN to be oxidized and the TKN to be used for synthesis.
4. Calculate the alkalinity to be consumed during nitrification and produced during denitrification. Note that if a residual alkalinity of approximately 100 mg/L as $CaCO_3$ can be maintained, a suitable pH for nitrification will exist in the reactor.
5. For the temperature, pH, dissolved oxygen and ammonia level in the reactor, calculate μ_n and θ_c^m from Equations 4.21 and 4.24.
6. Select a safety factor and calculate θ_c^d from Equation 4.25.
7. From θ_c^d, calculate μ_h as shown in Example 4.1.
8. Calculate the reactor volume and hydraulic retention time for nitrification as shown in Example 4.1.
9. Select a denitrification rate from pilot studies or from the literature (see Table 4.2).

10. Using the denitrification rate and MLVSS concentration, determine the additional reactor volume required for denitrification as shown in Example 4.2.
11. Calculate oxygen requirements and select aerators. Aerator placement will be selected to ensure anoxic zones.

Though there are considerable data available in the literature regarding kinetic constants which may be used in the design of oxidation ditches, there is no substitute for pilot studies. Each wastewater is different. The money spent on pilot studies is often recovered many times over in the design savings based on information obtained during pilot studies. Pilot studies are always recommended if the wastewater is available.

Oxygen Supply

The oxygen required at waste conditions in the oxidation ditch may be calculated in a variety of ways [1,4-6]. One approach to the calculation of oxygen requirements is to consider that all of the BOD_L is oxidized except for that fraction used for synthesis. In the same manner all nitrogen is oxidized except for that used for synthesis, and oxygen is available from denitrification. This may be stated as:

O_2 required = BOD_L removed – BOD_L of sludge wasted + oxygen demand of

NH_4-N removed – oxygen demand of NH_4-N in sludge wasted –

$$\text{oxygen available from denitrification} \tag{9.1}$$

Noting that the ultimate BOD of the VSS is usually taken as 1.42 times the VSS concentration; that the oxygen demand for nitrification may be assumed to be 4.5 times the NH_3-N concentration; that the VSS may be assumed to be approximately 12.4% nitrogen; and that the oxygen available from denitrification is approximately 2.6 mg O_2/mg NO_3-N reduced; we can rewrite Equation 9.1 as:

$$O_2 \text{ req.} = Q \left[\frac{(S_0 - S)\, 8.34}{1 - e^{-kt}} - 1.42\, P_x \left(\frac{VSS}{TSS}\right) + 4.5\, (N_0\text{-}N) \times 8.34 \right.$$

$$\left. - 0.56\, P_x \left(\frac{VSS}{TSS}\right) - 2.6\ \Delta NO_3 \times 8.34 \right] \tag{9.2}$$

where O_2 req. = oxygen required, lb/day
       Q = plant flow rate, MGD

S_o = influent BOD_5, mg/L
S = effluent soluble BOD_5, mg/L
k = BOD rate constant, day^{-1}
t = time of BOD test = 5 days
P_x = sludge wasted, lb/day
VSS/TSS = fraction of volatile solids in sludge
N_o = influent ammonia, mg/L NH_3-N
ΔNO_3 = NO_3-N reduced, mg/L NO_3-N

Equation 9.2 may be used to calculate the oxygen demand. The actual oxygen demand must then be converted to standard conditions to select the aerators. This is discussed in Chapter 5.

In a system designed for denitrification, the aerators must be placed in such a way as to ensure the maintenance of an anoxic zone. By dividing the oxygen required per unit time by reactor volume, an oxygen uptake rate in terms of oxygen used per unit volume may be determined. By calculation of oxygen input at each aerator and of the oxygen uptake rate, the length of ditch required to reach the anoxic zone may be determined. This will be demonstrated in Example 9.1.

Clarifier Design

The success of any activated sludge treatment system depends on the ability to separate and thicken the biological solids. A vast majority of the poor performances of activated sludge systems may be attributed to failure of solids separation or concentration. In oxidation ditch treatment systems, the clarifier must perform the function of solids separation as well as solids concentration.

Ideally, pilot studies will be carried out and sufficient data will be obtained to permit clarifier sizing on the basis of solids flux [4] analysis or other theoretical methods. In many cases, however, it is necessary to rely on literature values for overflow rate, solids loading rate and weir loading rate. Suggested values from the literature [1,4-9] have been summarized in Table 9.1.

Table 9.1. Recommended Clarifier Design Parameters

Process	Overflow Rate, gpd/ft^2	Solids Loading Rate, lb DS/ft^2/day	Weir Loading Rate, gpd/ft
Extended Aeration	300-500	4-20	10,000-15,000
Conventional	400-800	15-30	10,000-15,000

The return sludge rate required to maintain a desired MLVSS concentration in the oxidation ditch may be calculated from a simple mass balance around the ditch if the return sludge concentration is known. The mass balance yields:

$$QX_o + Q_R X_R = (Q + Q_R)X \qquad (9.3)$$

where Q = plant flow, MGD
Q_R = return activated sludge flow, MGD
X_o = influent VSS concentration, mg/L
X_R = return activated sludge VSS concentration, mg/L
X = oxidation ditch MLVSS concentration, mg/L

In Equation 9.3, total suspended solids may be used in place of volatile suspended solids as long as either total or volatile solids are used throughout the equation.

Sludge

Although the kinetic design will allow calculation of biological sludge production, account should be taken of the inert material present and the solids loss over the weirs of the clarifier. One approach to this is:

$$P_X = \left[Q\Delta S \left(\frac{Y}{1 + k_d \theta_c} \right) + X_I Q - X_e Q \right] 8.34 \qquad (9.4)$$

where P_X = total waste sludge, lb/day
Q = plant flow, MGD
ΔS = influent BOD_5 - effluent soluble BOD_5 mg/L
Y = growth yield coefficient, lb VSS produced/lb BOD_5 removed
k_d = endogenous decay coefficient, day^{-1}
θ_c = solids retention time, days
X_I = inert fraction of influent suspended solids (influent TSS – influent VSS), mg/L
X_e = effluent TSS, mg/L

When the oxidation ditch is run in the conventional mode, the waste activated sludge produced will be unstable and will require stabilization prior to disposal. When the oxidation ditch is operated in the extended aeration mode, sludge production will be low and the sludge will be stabilized.

The most common sludge handling methods for the stabilized sludges produced from oxidation ditches are drying on sand drying beds and landspreading or landfill of the dried sludge.

Design Example

In order to demonstrate the design of oxidation ditches, a design example will be carried out. No attempt has been made to design a complete wastewater treatment plant. It must be recognized that design is an iterative process and that solids balances, effects of recycle streams and the like must be considered in the real design process.

Example 9.1. Design an oxidation ditch plant to treat the following domestic wastewater:

	alkalinity	=	280 mg/L as $CaCO_3$
Q = 3 MGD	peak to average		
BOD_5 = 220 mg/L	flow	=	1.5
TSS = 240 mg/L	minimum		
VSS = 180 mg/L	temperature	=	$15°C$
TKN = 35 mg/L	maximum		
	temperature	=	$25°C$

The effluent requirements are:

BOD_5 = 20 mg/L
TSS = 20 mg/L
NH_4-N = 2 mg/L
NO_3-N = 10 mg/L
sludge to be stabilized

The first step is to select the design parameters to be used. To ensure that the sludge is stabilized, a minimum solids retention time of 30 days will be used. The mixed liquor will be assumed to be 70% volatile. A total MLSS concentration of 4000 mg/L will be selected to allow for higher solids concentrations in response to peak loads. The reactor type to be used will be the Carrousel system. The following parameters will be used in the design of the aeration system for oxidation ditch:

Aerator type: low-speed vertical turbine aerators
Aerator efficiency: 3.0 lb O_2/hp-hr
Reactor DO: 2.0 mg/L
Oxygen for nitrification: 4.5 mg O_2/mg NH_4-N oxidized
Oxygen available: 2.6 mg O_2/mg NO_3-N reduced
α: 0.90
β: 0.98

Kinetic constants:

$$Y_H = 0.60 \ \frac{mg \ VSS}{mg \ BOD_5} \quad \text{(from Table 2.3)}$$

$K_d = 0.05 \text{ day}^{-1}$ (Table 2.3)

$$q_{dn} = 0.02 \ \frac{\text{lb NO}_3\text{-N}}{\text{lb MLVSS-d}} \ @ 20°C \quad \text{(from Table 4.1)}$$

$k_1 = 0.23 \text{ day}^{-1}$

$K_{O_2} = 1.3 \text{ mg/L}$

Residual alkalinity: 100 mg/L as $CaCO_3$ to maintain pH $\geqslant 7.2$
Alkalinity required: 7.14 mg ALK/mg NH_4-N oxidized
Alkalinity produced: 3.0 mg ALK/mg NO_3-N reduced
Safety factor for nitrification: 2.5
Temperature correction factor for denitrification, θ: 1.08

The clarifier design parameters are:

Overflow rate: 300 gpd/ft^2 (Table 9.1)
Solids loading rate: 10 lb/ft^2/day (Table 9.1)
Weir loading rate: 15,000 gpd/ft (Table 9.1)
Clarifier type: center feed suction; clarifier with scraper mechanism
Minimum number of clarifiers: 2
Return sludge concentration: 10,000 mg/L

The first step will be to determine soluble BOD_5 required in the effluent to achieve 20 mg/L BOD_5 effluent (see Example 2.3).

$$\text{soluble BOD}_5 = 20 \text{ mg/L} - 0.7 \times 20 \text{ mg/L} \times 1.42 (1 - e^{-0.23 \times 5})$$

$$\text{soluble BOD}_5 = 6.4 \text{ mg/L}$$

Now, assuming that $\theta_c = 30$ day (this must be checked) we can calculate the yield of biological solids:

$$\text{biological solids produced} = \left(\frac{Y}{1 + k_d \theta_c} \right) Q \Delta S$$

$$= \left(\frac{0.6}{1 + 0.05 \times 30} \right) \times 3 \times (220 - 6) \times 8.34 = 1285 \text{ lb/day}$$

The fraction of TKN used for synthesis may be approximated by assuming that the biological solids are approximately 12.4% nitrogen. Thus, the amount of nitrogen used daily for synthesis is:

$$\text{TKN for synthesis} = 0.124 \times 1285 \text{ lb/day} = 159.3 \text{ lb/day}$$

This represents:

$$\frac{159.3}{3 \times 8.34} = 6.4 \text{ mg/L of TKN in the influent used for synthesis}$$

Thus, the nitrogen which must be oxidized:

$$NH_3\text{-N oxidized} = 35 \text{ mg/L} - 6.4 \text{ mg/L} - 2 \text{ mg/L} = 26.6 \text{ mg/L}$$

The amount of denitrification required is then:

$$\text{denitrification required} = 26.6 \text{ mg/L} - 10 \text{ mg/L} = 16.6 \text{ mg/L } NO_3\text{-N}$$

We now can perform an alkalinity balance. Alkalinity produced during carbonaceous BOD removal, from Figure 4.1, is approximately 0.1 mg alkalinity produced per mg BOD_5 removed.

$$\text{residual alkalinity} = 280 \text{ mg/L} - 7.14 \times 26.6 \text{ mg/L} + 3.0 \times 16.6 \text{ mg/L}$$

$$+ 0.1 \times 214 \text{ mg/L} = 161.3 \text{ mg/L as } CaCO_3$$

which is sufficient to maintain the pH at or above 7.2 mg/L

We can now determine the growth rate for nitrification:

$$\mu_n = \left[0.47 e^{0.098T - 15} \right] \left[\frac{2}{2 + 10^{0.05 \times 15 - 1.158}} \right] \times \left[\frac{2}{2 + 1.3} \right]$$

$$\mu_n = 0.237 \text{ day}^{-1}$$

Thus:

$$\theta_c^m = \frac{1}{0.237} = 4.2 \text{ days}$$

Using a safety factor of 2.5, the design θ_c = 10.5 days. We have selected a 30-day θ_c to ensure stabilized sludge. Therefore, we will calculate growth rates based on a 30-day solids retention time.

$$\mu_n = \mu_h = \frac{1}{30} = 0.033 \ \text{day}^{-1}$$

We will now calculate specific substrate utilization (see Example 4.1):

$$U = \frac{\mu_n + k_d}{Y} = \frac{0.033 + 0.05}{0.6} = 0.139 \ \frac{\text{lb BOD}_5}{\text{lb MLVSS-day}}$$

Using the specific substrate utilization rate and the design value for MLSS of 4000 mg/L, we can now determine the volume of the ditch and the hydraulic retention time.

The MLVSS concentration:

$$\text{MLVSS} = 0.7 \times 4000 = 2800 \ \text{mg/L}$$

The MLVSS required:

$$\text{MLVSS required} = \frac{214 \times 3 \times 8.34 \ \text{lb BOD rem./day}}{0.139 \ \dfrac{\text{lb BOD rem.}}{\text{lb MLVSS-day}}} = 38,520 \ \text{lb}$$

Volume of the reactor:

$$V = \frac{38,520 \ \text{lb MLVSS}}{8.34 \times 2800 \ \text{mg/L}} = 1.65 \ \text{MG}$$

Hydraulic retention time:

$$\theta = \frac{V}{Q} = \frac{1.65 \ \text{MG}}{3 \ \text{MGD}} \times 24 \ \text{hr/day} = 13.2 \ \text{hr}$$

Now we can calculate the volume and retention time required for denitrification.

At 15°C:

$$q_{dn} = 0.020 \times 1.08^{-5} = 0.0136 \quad \frac{\text{lb } NO_3\text{-N red.}}{\text{lb MLVSS-day}}$$

The NO_3-N to be reduced:

$$NO_3\text{-N reduced} = 16.6 \text{ mg/L} \times 3 \times 8.34 = 415.3 \text{ lb/day}$$

The MLVSS required for denitrification:

$$\text{MLVSS} = \frac{415.3 \text{ lb/day}}{0.0136 \text{ lb/lb-day}} = 30{,}539 \text{ lb MLVSS}$$

Volume required for denitrification:

$$V = \frac{30{,}539 \text{ lb/day}}{2800 \text{ mg/L} \times 8.34} = 1.31 \text{ MG}$$

The hydraulic retention time for denitrification:

$$\theta = \frac{1.31 \text{ MG}}{3 \text{ MGD}} \times 24 \text{ hr/day} = 10.5 \text{ hr}$$

Thus the overall oxidation ditch will have a volume of 2.96 MG and a hydraulic retention time of 23.7 hr.

For the Carrousel reactor a depth of 11 ft will be used along with a width of 24 ft. The length of reactor required is:

$$L = \frac{2.96 \times 10^6 \text{ gal}}{7.48 \text{ gal/ft}^3 \times 24 \text{ ft} \times 11 \text{ ft}} = 1500 \text{ ft}$$

The aerobic length is 836 ft and the anoxic length is 664 ft. The layout used is shown in Figure 9.1.

We will now size the aerators. The actual oxygen requirement may be calculated:

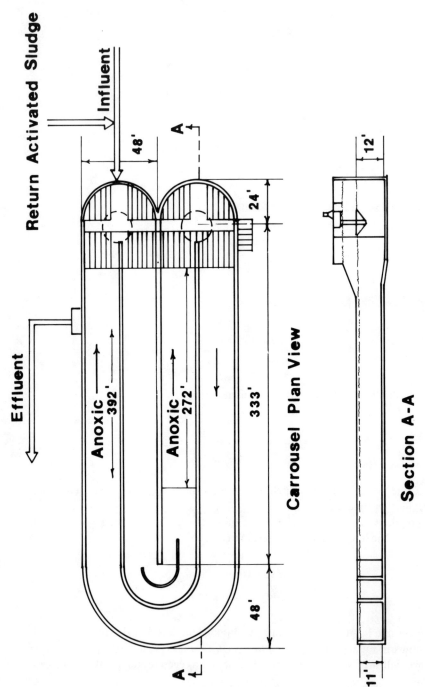

Figure 9.1. Oxidation ditch layout for Example 9.1.

$$AOR = \frac{214 \text{ mg/L} \times 3 \text{ MGD} \times 8.34}{(1 - e^{-0.23 \ X5})} - 1.42 \times 1285 \text{ lb}$$

$$\underbrace{}_{\text{BOD removed}} \qquad \underbrace{}_{\substack{\text{BOD of volatile solids} \\ \text{produced}}}$$

$$+ \underbrace{4.5 \times 26.6 \text{ mg/L} \times 3 \text{ MGD} \times 8.34}_{\text{oxygen required for nitrification}} - \underbrace{2.6 \times 16.6 \text{ mg/L} \times 3 \text{ MGD} \times 8.34}_{\text{oxygen available from denitrification}}$$

$$AOR = 7926 \text{ lb/day}$$

Converting to standard conditions:

$$SOR = \frac{7926 \times 9.07}{0.9 \ (8.08 - 2)1.024^5}$$

$$SOR = 11,669 \text{ lb/day} \times \frac{1 \text{ day}}{24 \text{ hr}} = 486 \text{ lb/day}$$

At 3.0 lb/hp-hr:

$$hp \text{ required} = 162 \text{ total hp}$$

Therefore, two 100-hp low-speed aerators would meet the design requirements.

Now calculate return activated sludge requirements and total waste sludge. Return activated sludge, calculating on a total solids basis:

$$240 \text{ mg/L} \times 3 \text{ MGD} + 10,000 \text{ mg/L} \times Q_R = (3 + Q_R) \text{MGD} \times 4000 \text{ mg/L}$$

$$Q_R = 1.88 \text{ MGD}$$

Waste sludge:

$$WAS = 1285/0.7 + (240 - 180) \times 3 \times 8.34 = 3337 \text{ lb dry solids per day}$$

If the sludge is wasted from the RAS line, the volume of WAS:

$$\text{volume}_{WAS} = \frac{3337 \text{ lb/day}}{0.01 \times 8.34} = 40,012 \text{ gpd}$$

Now size the clarifiers. Based on overflow rate:

$$\text{area required} = \frac{3 \times 10^6 \text{ gpd}}{300 \text{ gpd/ft}^2} = 10,000 \text{ ft}^2$$

Based on solids loading:

$$\text{solids to clarifier} = (3 + 1.88)\text{MGD} \times 4000 \text{ mg/L} \times 8.34 = 162,797 \text{ lb/day}$$

$$\text{area required} = \frac{162,797 \text{ lb/day}}{10 \text{ lb/ft}^2\text{/day}} = 16,280 \text{ ft}^2$$

Therefore solids loading controls:

$$\text{area required} = 16,280 \text{ ft}^2$$

Using two clarifiers:

$$\text{area each} = 8140 \text{ ft}^2$$

$$\text{diameter required} = \left[\frac{8140 \times 4}{\pi} \right]^{\frac{1}{2}} = 101.8 \text{ ft}$$

Therefore, two 100-ft-diameter clarifiers would meet the design requirements.

Check the weir length with inboard launders:

$$\text{weir length} = 2 \times \pi \times 100 = 628 \text{ ft}$$

$$\text{weir loading} = \frac{3 \times 10^6 \text{ gpd}}{628} = 4777 \text{ gpd/ft}$$

Design Summary Example 9.1

Oxidation Ditch

>Carrousel system
>Reactor volume: 2.96 MG
>Hydraulic retention time: 23.7 hr
>Solids retention time: 30 days
>SOR: 486 lb/hr
>Aerators: two 100-hp low-speed aerators
>Configuration: See Figure 9.1

Clarifiers

>Number: 2
>Type: center feed, suction with scrapers
>Size: 100-ft diameter, 15-ft side water depth, inboard weirs
>Return activated sludge: 1.88 MGD
>Waste activated sludge: 40,012 gpd

REFERENCES

1. "Wastewater Treatment Plant Design," Water Pollution Control Federation, MOP8 (1977).
2. "Design of Wastewater and Stormwater Pumping Stations," Water Pollution Control Federation, MOP No. FD-4 (1980).
3. Metcalf and Eddy, Inc., *Wastewater Engineering Collection and Pumping of Wastewater* (New York: McGraw-Hill Book Company, 1981).
4. Metcalf and Eddy, Inc., *Wastewater Engineering*, 2nd Edition (New York: McGraw-Hill Book Company, 1979).
5. Benefield, L. D., and C. W. Randall, *Biological Process Design for Waste-Water Treatment* (Englewood Cliffs, NJ: Prentice-Hall, 1980).
6. Sundstrom, D. W., and H. E. Klei, *Wastewater Treatment* (Englewood Cliffs, NJ: Prentice-Hall, 1979).
7. Schroeder, E. D., *Wastewater and Wastewater Treatment* (New York: McGraw-Hill Book Company, 1977).
8. Eckenfelder, W. W., Jr., "Cost Effective Advanced Biological Treatment Systems," Presented at Annual Meeting Nevada Water Pollution Control Association (Dec., 1977).

CHAPTER 10

ECONOMIC CONSIDERATIONS

It must be recognized that any discussion of economics of treatment processes must be general. Many factors affect the cost of any treatment facility. These factors may be considered to be process-, site-, region- and climate-specific. When generalizing a discussion of economics, we can discuss and compare process-specific costs with relative ease. Region- or climate-specific costs are more difficult to handle, but may be approached in a somewhat generalized fashion. Site-specific costs are virtually impossible to deal with in any general fashion.

Beyond the scope of this discussion, but of major importance in the designer's decision-making process, is the interrelationship of liquid and solids handling processes. There exist many examples of cost comparisons in the literature [1-5] which attempt to address this interrelationship. It is clear, for example, that an oxidation ditch being operated in the extended aeration range will have totally different requirements and costs for sludge treatment and disposal than will an oxidation ditch being operated in the conventional range.

TYPES OF COSTS

Before discussing the costs of oxidation ditches and competing processes, we should address the types of costs encountered.

Process-Specific Costs

It is clear that each type of process will have its own requirements. For example, a plant utilizing a high-rate activated sludge process will have a

157

much smaller volume requirement for its aeration basin and a much lower oxygen requirement than will a plant utilizing extended aeration; smaller volume and lower oxygen requirements mean lower costs. Site-specific and region-specific costs will affect the actual costs, but in this case the relative cost of aeration basins and equipment may be accurately described by comparing the average cost of each process.

Region- or Climate-Specific Costs

Regional differences, primarily climate-related, will have significant impact on the construction costs of treatment facilities. Climate-related factors will often dictate process or equipment selection. Ammonia stripping for nitrogen removal may be seriously considered in warm climates but quickly eliminated from consideration in cold climates. Submerged rather than surface aeration may be favored in cold climates due to surface cooling and ice formation with surface aerators.

Climate affects the costs of plants in many ways. An oxidation ditch designed for nitrogen removal at a minimum temperature of $5^{\circ}C$ will be far larger, and more expensive, than one designed to meet the same effluent requirements at a minimum temperature of $15^{\circ}C$. Plant architecture will be different in different regions. The size of oxygen transfer equipment, clarifiers and sand drying beds are all temperature- or climate-dependent.

In general, cost curve comparisons, as presented in this chapter, can give reasonably accurate relative construction costs of various processes despite regional differences. Absolute or actual costs and cost differences cannot be accurately obtained from cost curves. Because of regional differences it is important, in using cost curves for different processes, to understand temperature effects on design parameters for different processes, and to allow for these effects in interpreting cost curve data.

Labor costs vary significantly with location in this country and around the world. Both construction and operating costs will vary significantly with location. Fuel, power, chemical and other operating costs also will vary regionally. These differences cannot be accurately described in general cost curves.

Site-Specific Costs

There are many site-specific factors which will affect the cost of various processes. The cost of land will affect the economics of different processes. Soil conditions, height of water table and other subsurface conditions will have significant impact on the cost of different processes. The plant site itself and its location relative to built-up areas may determine the need for expensive architectural treatment, buffer zones, and noise and odor control facilities. These costs cannot be reflected in general cost curves.

COST DATA

The cost data for oxidation ditches and other competing processes presented are taken from work carried out by EPA [6]. All costs given are referenced to EPA treatment plant index of 262.3, Fall, 1976. Costs given for construction include building, laboratory and sludge drying beds, but do not include land, engineering, legal, administrative and financing costs.

In utilizing the data presented it must be remembered that these costs are based on specific design assumptions for municipal wastes. The reader is referred to the original work [6] for these design assumptions. Care must be taken in using these cost comparisons for wastes and design parameters other than those assumed.

Construction Costs

In oxidation ditches, the construction costs of carbonaceous BOD removal with or without nitrification are considered to be essentially the same [6]. While not strictly true, this assumption is valid within the accuracy of the cost curves. Construction costs for oxidation ditches and competing processes for BOD removal are presented in Table 10.1. These costs are presented graphically in Figure 10.1.

It may be seen from Table 10.1 and Figure 10.1 that in the ranges for which each process is applicable, and with the assumed design parameters, no significant difference in construction cost exists between competing processes. Differences in process performance (discussed in Chapter 8) as well as operation and maintenance costs must be considered.

Table 10.1. Construction Costs for BOD Removal ($1000) [6]

Process	Plant Capacity, MGD					
	0.05	0.1	0.5	1.0	5.0	10.0
Oxidation Ditch	170	195	330	600	2000	3350
Contact Stabilization		165	320	475		
Extended Aeration	108	180	390			
Conventional Activated Sludge				1045	2645	4138

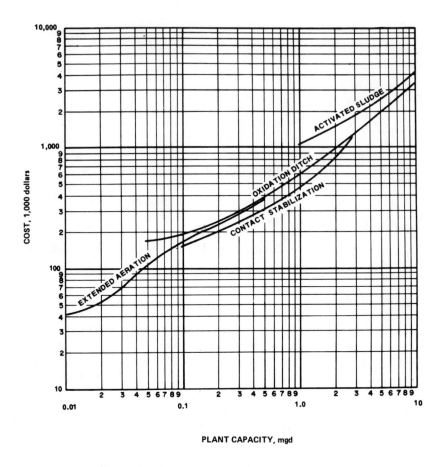

Figure 10.1. Construction costs for BOD removal [6].

When nitrification is considered, no change in the construction cost for extended aeration or oxidation ditch plants is expected. Activated sludge plants, however, will require considerable additional aeration basin capacity as well as other modifications. Contact stabilization does not lend itself to nitrification processes. The construction costs for nitrification systems are presented in Table 10.2 for the range of flows applicable to both oxidation and activated sludge plants. When nitrification is considered, it is clear from Table 10.2 that the construction cost of oxidation ditches is significantly less than for conventional activated sludge with nitrification. At 1 MGD the cost of an oxidation ditch is less than half that of a single-stage activated sludge plant producing a nitrified effluent.

When denitrification is required, the construction cost advantage demonstrated by oxidation ditches is even more apparent. Table 10.3 presents the construction costs for plants requiring nitrogen removal using oxidation ditches and activated sludge with both mixed slurry and fixed film reactors. The costs shown in Table 10.3 for activated sludge processes with denitrification are based on the use of separate denitrification facilities and the use of methanol as the carbon source. Recent work in denitrification has resulted in a reduction in these costs. Thus, Table 10.3 somewhat overstates the costs for denitrification in the activated sludge process. However, the cost of denitrification in oxidation ditches remains less than with activated sludge processes.

Operation and Maintenance Costs

In discussing operation and maintenance costs, the same cautions and restrictions previously presented for construction costs apply. The O&M costs for carbonaceous BOD removal are presented in Table 10.4.

In a similar manner as for construction costs the O&M costs for nitrification and for denitrification are presented in Tables 10.5 and 10.6. As pre-

Table 10.2. Construction Costs for BOD Removal and Nitrification ($1000) [6]

Flow, MGD	Oxidation Ditch	Single-Stage Activated Sludge	Two-Stage Activated Sludge
1	600	1240	1490
5	2000	3283	3942
10	3350	5207	6171

Table 10.3. Construction Costs for BOD Removal
Nitrification and Denitrification ($1000) [6]

Flow, MGD	Oxidation Ditch	Activated Sludge Mixed Slurry Denitrification		Activated Sludge Fixed Film Denitrification	
		Single-Stage	Two-Stage	Single-Stage	Two-Stage
1	629	1834	2084	1893	2143
5	2100	4811	5470	4521	5180
10	3534	7795	8759	7562	8526

Table 10.4. O&M Costs for Carbonaceous BOD Removal ($1000/yr) [6]

Process	Plant Capacity, MGD					
	0.05	0.1	0.5	1.0	5.0	10.0
Oxidation Ditch	19	22	42	52	230	460
Contact Stabilization		21	47	98		
Extended Aeration	13	23	55			
Conventional Activated Sludge				82	195	310

Table 10.5. O&M Costs for BOD Removal and Nitrification ($1000/yr) [6]

Flow, MGD	Oxidation Ditch	Single-Stage Activated Sludge	Two-Stage Activated Sludge
1	52.6	90	104
5	231	227	251
10	461	365	

Table 10.6. O&M Costs for BOD Removal Nitrification and Denitrification ($1000/yr) [6]

Flow, MGD	Oxidation Ditch	Single-Stage Activated Sludge Mixed Slurry Denitrification	Single-Stage Activated Sludge Fixed Film Denitrification
1	57	143	140
5	230	355	340
10	460	620	560

viously stated, the costs for denitrification activated sludge systems are overstated as the result of recent improvements in denitrification technology.

Overall Costs

When the capital cost is amortized for 20 years at 7%, total annual costs may be determined. It is recognized that 7% interest is low in today's market. However, since little difference exists in the capital cost of the processes, the low interest rate should not greatly distort the comparisons.

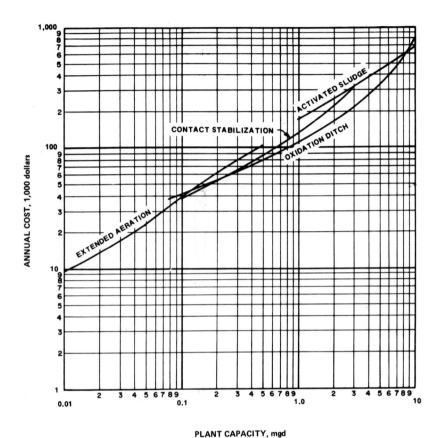

Figure 10.2. Total annual cost of carbonaceous BOD removal [6].

Table 10.7. Total Annual Cost, BOD Removal, Nitrification and Denitrification ($1000/yr) [6]

Flow, MGD	Oxidation Ditch	Single-Stage Activated Sludge Mixed Slurry Denitrification	Single-Stage Activated Sludge Fixed Film Denitrification
1	117	305	325
5	343	795	755
10	815	1370	1040

Figure 10.2 presents the total annual cost comparison for carbonaceous BOD removal. A total annual cost comparison for BOD removal, nitrification and denitrification is presented as Table 10.7.

SUMMARY

The economic data indicate that there is little difference in the cost of various processes for conventional BOD removal (Figure 10.2). While Figure 10.2 shows small cost differences between processes, these differences are not significant, given the assumptions required and regional- and site-specific cost differences.

When nitrification and denitrification are required, oxidation ditches show a significant cost advantage over competing processes. Although the data presented probably overstate denitrification costs for competing processes, the cost differences are so large that oxidation ditch plants show a clear cost advantage when nitrification and denitrification are required.

The design engineer responsible for process selection must recognize that economics is only one factor, albeit an important factor, in the decision-making process. Other factors, such as ease of operation and the ability to reliably meet the required effluent, also must be considered in process selection.

REFERENCES

1. Wilson, R. W., K. L. Murphy, P. M. Sutton and S. L. Lackey, "Design and Cost Comparison of Biological Nitrogen Removal Processes," *J. Water Poll. Control Fed.*, 1294-1302 (1981).
2. Guo, P. H. M., D. Thirumurthi and B. E. Jank, "Evaluation of Extended Aeration Activated Sludge Package Plants," *J. Water Poll. Control Fed.*, 53(1):33-42 (1981).
3. Bell, B. A., and T. M. Zaferatos, "Evaluation of Alternative Solids Handling Methods for Advanced Waste Treatment Lime Sludges," *J. Water Poll. Control Fed.*, 49(1):146 (1977).
4. Bell, B. A., "Energy Conservation and Production from the Anaerobic Digestion of Thermally Conditioned Sludges and Decant Liquors," Proceedings Energy Optimization of Water and Wastewater Management for Municipal and Industrial Applications Conference, U.S. DOE, ANL/EES-TM-96 (1979).
5. Stensel, H. D., and J. H. Scott, "Cost Effective Advanced Biological Treatment Systems," Presented at Annual Meeting Nevada Water Pollution Control Association (Dec., 1977).
6. "A Comparison of Oxidation Ditch Plants to Competing Processes for Secondary and Advanced Treatment of Municipal Wastes," U.S. EPA, EPA-600/2-78-051 (1978).